FUNDAMENTALS OF ECOLOGY LABORATORY MANUAL

Fred E. Smeins • Jason B. West • Rachel R. Fern
R. Douglas Slack • Michael L. Morrison

FOURTH EDITION

Kendall Hunt
publishing company

Kendall Hunt
publishing company

www.kendallhunt.com
Send all inquiries to:
4050 Westmark Drive
Dubuque, IA 52004-1840

ISBN 978-1-5249-0894-2

Published in the United States of America

Contents

Preface

The science of Ecology deals with the interrelationships of organisms and their environment. Knowledge of ecological concepts and their applications is essential to understanding the world in which we live. Ecology is a relatively old science and grew out of the natural history movement of the nineteenth century. The field is still developing and maturing and as a science matures it attempts to become more exacting and quantitative in its approach. A major objective of ecology is to more adequately describe and understand the structure and function of ecosystems. This laboratory manual is an introduction to the rudiments of natural ecosystem description and analysis.

This manual evolved out of a sophomore-level course in Fundamentals of Ecology taught at Texas A&M University to both majors and non-majors. It was developed primarily for use in a field-oriented course, although most exercises can be adapted to indoor situations. The manual is directed primarily at students in natural resource fields and the biological sciences.

Exercises are designed to introduce students to the concepts, procedures and equipment used in the field investigation of biotic and abiotic components of an ecosystem. Additionally, some exercises deal with functional and dynamic features of ecosystems such as plant and animal adaptations, ecological succession, and climatic, edaphic and biotic factor interactions, which combine to produce the ecosystems we observe in nature. Terrestrial ecology is emphasized but an introduction to aquatic ecology is included.

The fourth edition is restructured with chapters organized in eight sections following the Preface: Environmental Factors, Adaptation, Population Studies, Aquatic Ecosystem Studies, Community Sampling, Ecological Survey and Study Design, and Ecological Concepts and Modeling which is followed by an Appendix. Substantial revisions were made to all chapters, with an increased emphasis on restoration ecology in the final chapter. These changes will hopefully make the manual more applicable to other potential users.

The authors wish to acknowledge the contributions of many students, teaching assistants and colleagues in the development of this manual. We graciously accept their contributions and hope their ideas have been correctly and adequately incorporated. We especially thank Dr. X. Ben Wu for his insights and substantial efforts expended on earlier editions of this manual. Errors and inconsistencies are accepted as the responsibility of the authors. It is hoped that the manual will be useful and will fill needs beyond the immediate audience for which it was developed.

1

Natural Regions of Texas

Texas has a remarkable variety of climatic, geologic, topographic and edaphic conditions. This variation in the environment provides a diversity of habitats for a great number of plants and animals. For example, there are found within the State almost 5,000 species of vascular plants, approximately 530 bird species, in excess of 150 species of mammals and nearly 200 amphibians and reptiles. These plants and animals come together to form an almost infinite number of natural communities.

Environment

To appreciate the ecological diversity of the State, it is necessary to become familiar with some of the environmental variation that exists. Precipitation ranges from an annual average high of nearly 150 centimeters (60 inches) in the Pineywoods near the Louisiana border to a low of less than 25 centimeters (10 inches) near El Paso in the Trans-Pecos. Over most of the State, May and September are months of peak precipitation while mid-summer tends to be quite dry. Exceptions are the southeastern portions of the Pineywoods and the upper Coastal Plains which have a fairly uniform monthly rainfall and the Trans-Pecos which typically receives most of its limited precipitation during July and August. Droughts of a few months to a few years are possible in any part of the State but are particularly common in areas that receive less than 60 centimeters (25 inches) of annual precipitation. Monthly and annual precipitation for selected natural regions in Texas are presented in Table 1.1. Evaporation ranges from 250 centimeters (100 inches) in the southwest to 115 centimeters (45 inches) in East Texas.

The average annual frost-free period, or growing season, varies from 180 days in the northwestern corner of the state to over 330 days in parts of South Texas. Annual mean temperatures vary from 12°C (54°F) in the northwest to 23°C (74°F) in South Texas. During winter months, freezing temperatures can occur in any part

TABLE 1.1 Monthly and Annual Precipitation Normals (Centimeters) for Selected Natural Regions (Adapted from Carr 1969)

Natural Region	Jan.	Feb.	Mar.	Apr.	May.	June.	July.	Aug.	Sept.	Oct.	Nov.	Dec,	Annual
High Plains	1.7	1.5	1.8	3.5	7.7	6.2	6.1	5.4	5.0	4.6	1.5	1.9	46.9
Pineywoods	10.7	9.7	9.1	11.6	13.0	8.5	8.7	7.4	7.5	8.0	10.8	11.7	116.7
Trans-Pecos	1.7	1.0	1.2	1.4	3.2	3.4	4.6	4.2	4.4	3.0	1.0	1.4	30.5
Edwards Plateau	4.1	4.0	3.7	6.1	9.0	6.8	5.7	5.1	7.9	5.9	3.3	4.2	65.8
South Texas	3.1	3.0	2.4	4.8	7.9	6.5	4.5	5.4	8.0	5.3	2.5	3.2	56.6

of the State and periodic cold fronts or "blue northers" can produce severe freezes. Summers are usually hot and in most regions, particularly in the western part of the State, windy conditions are prevalent.

Physiographically the landscapes of Texas can generally be divided into three types: 1) relatively level plains of the High and Coastal Plains, 2) basins and mountains of the Trans-Pecos, and 3) rolling topography of Central, South and East Texas. Elevations vary from sea level along the Gulf Coast to over 2650 meters (8751 feet) on Guadalupe Peak near the New Mexico border in the Trans-Pecos. Generally, the State slopes from northwest to southeast as illustrated by the drainage systems that cross the State.

Geologically there are outcrops of many different ages and with much physical and chemical variation. There are pre-Cambrian granites (e.g., Llano Uplift), Cretaceous limestones (e.g., Edwards Plateau, Blackland Prairies), Cenozoic sands and clays (e.g., Oak Woods and Prairies, High Plains) and Recent alluvial deposits along streams and rivers. Most of the surficial deposits were produced as sediments from epicontinental seas that have variously covered the state through geologic time; however, volcanic activity, folding, faulting and other geologic phenomena have also contributed to the State's geologic features.

The geologic substrates (parent material), as influenced by climate, relief, organisms and time have produced the many soils of the State. These soils, with their great variation in physical (texture, structure, depth, etc.) and chemical (nutrient content, salinity, etc.) properties, provide the edaphic conditions where terrestrial plant and animal communities find favorable environments for their existence. Soils vary from deep sands of the Coastal Sand Plains to heavy clay soils of the Blackland and Coastal Prairies with nearly every possible combination in between.

In addition to the physical environment, all organisms exist in the presence of other organisms and thus, part of their environment includes biotic interactions that occur within a community. Competition, predation, parasitism and other biotic interactions can greatly affect the structure and composition of natural communities. Biotic influences are often the result of man's activities. Man has been part of the Texas environment for at least 12,000 years and possibly much longer. All through that period of time man has had an impact on natural systems, but as modern man with his greater technology and larger population arrived on the scene, the impact became greater and more permanent. Man has greatly altered Texas' natural systems by control of wildlife populations, lumbering, cultivation, burning, livestock grazing, urban development and other activities.

In summary, the natural ecosystems of Texas are controlled generally by the macroclimatic and soil-topographic conditions that exist in an area. These factors, often called site or habitat factors, set the stage upon which organisms arrive, become established, interact with one another and sort themselves into communities. These communities can then be variously impacted and altered by man's activities. Some of these impacts have been detrimental: species have been driven to extinction [e.g., red wolf (Canis rufus), ivory-billed woodpecker (Campephilus principalis)], species have been greatly reduced in numbers [e.g., Attwater's prairie-chicken (Tympanuchus cupide), whooping crane, Texas wildrice (Zizania texana)], habitats have been greatly altered or diminished in extent (e.g., wetlands, bottomland forests, tall grass prairies) and soil erosion accelerated to a point where the rate of loss exceeds rate of development by a factor of 2 to 4 times. While many impacts are negative, man has the capability to renew, maintain and prevent severe disturbance and destruction of our natural ecosystems. These ecosystems are renewable and with an adequate knowledge and understanding of their ecological characteristics, man can not only derive benefits from them but also maintain them in an ecologically healthy and productive state.

Natural Regions

The following sections present an overview of the ecological characteristics of the Natural Regions of Texas (Figure 1.1). A brief description of the location, climate, soils and any unique environmental features is given in addition to lists of selected vascular plants, mammals, birds, amphibians and reptiles. Where possible, plant species are divided into generalized habitats and successional status. It must be recognized that often these distinctions are arbitrary since many species have broad ecological amplitudes and may occur in many habitats. Also, a particular species may be late successional on one soil within a region and successional on another soil. In some cases, an intermediate list of plants is given of those species that occur throughout many successional stages. For the Oak Woodlands (Post Oak Savannah) a somewhat more detailed presentation is given of vegetation-environmental relationships and successional patterns.

Animal lists are somewhat more difficult to define for each region. Because of their great mobility and the need of many animals to utilize a variety of habitats for their requirements, many species tend to occur across several regions. Thus, animal lists are only presented for four geographic extremes of the State, the Pineywoods (forests), the South Texas Plains (shrublands), the Coastal Prairies and Marshes (grasslands and wetlands) and the Trans-Pecos region (desert and mountains). Other regions tend to be mixtures of species that overlap from these regions or are species with a more or less cosmopolitan distribution across Texas. Examples of animals that have a wide distribution across the State are:

Mammals

Opossum	*Didelphis virginiana*
Red Bat	*Lasiurus borealis*
Raccoon	*Procyon lotor*

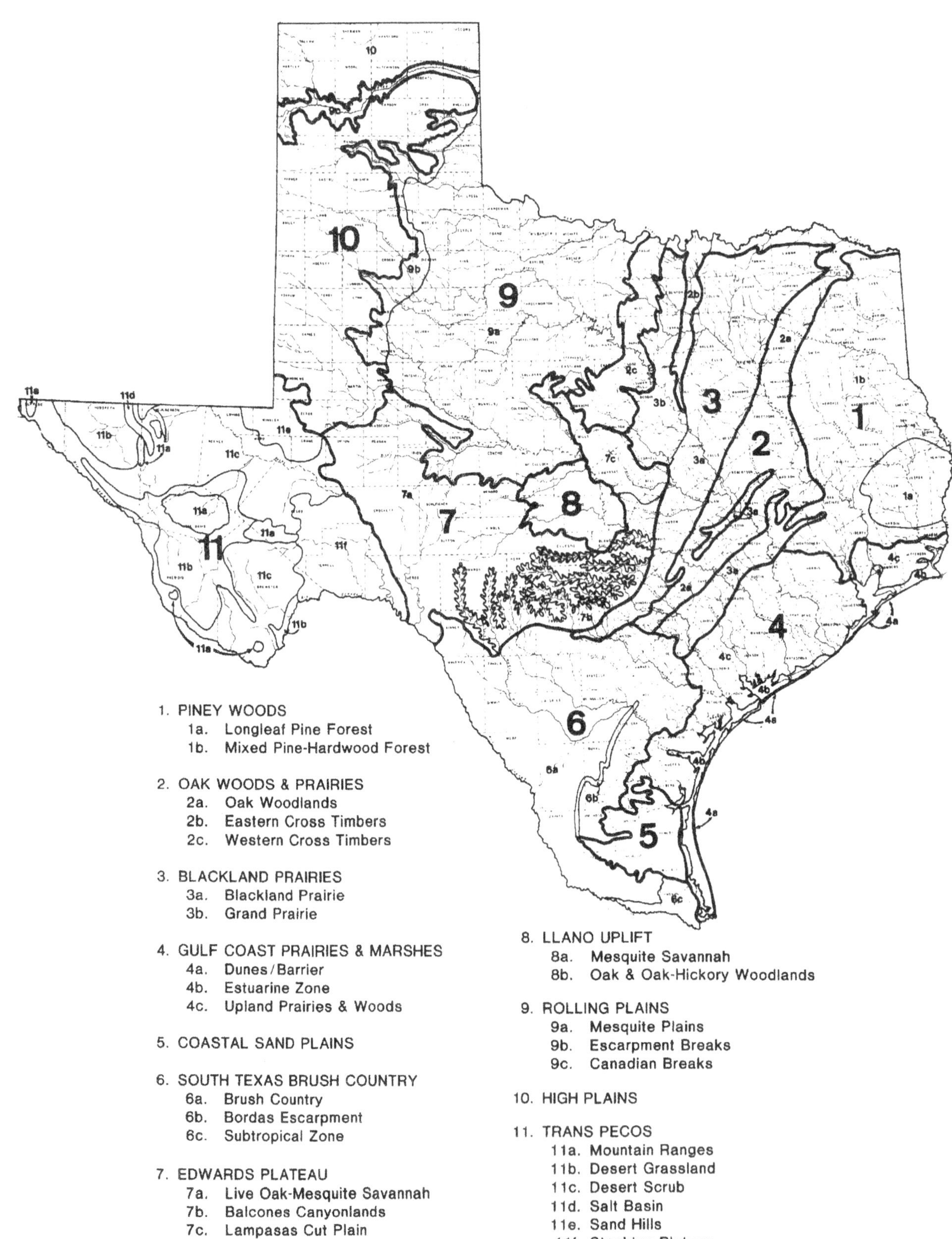

FIGURE 1.1 Natural Regions of Texas (adapted from Gould 1975 and LBJ School of Public Affairs 1978).

Striped Skunk	*Mephitis mephitis*
Gray Fox	*Urocyon cinereoargentus*
Coyote	*Canis latrans*
Bobcat	*Lynx rufus*
Hispid Pocket Mouse	*Perognathus hispidus*
White-footed Mouse	*Peromyscus leucopus*
Hispid Cotton Rat	*Sigmodon hispidus*
California Jackrabbit	*Lepus californicus*
Eastern Cottontail	*Sylvilagus floridanus*
White-tailed Deer	*Odocoileus virginianus*

Birds

Pied-billed Grebe	*Podilymbus podiceps*
Turkey Vulture	*Cathartes aura*
Northern Harrier	*Circus cyaneus*
Red-tailed Hawk	*Buteo jamaicensis*
American Kestrel	*Falco sparverius*
Great Blue Heron	*Ardea herodias*
Killdeer	*Charadrius vociferus*
Mourning Dove	*Zenaida macroura*
Greater Roadrunner	*Geococcyx californianus*
Great Horned Owl	*Bubo virginianus*
Common Nighthawk	*Chordeiles minor*
Belted Kingfisher	*Megaceryle alcyon*
Downy Woodpecker	*Picoides pubescens*
Horned Lark	*Eremophila alpestris*
Barn Swallow	*Hirundo rustica*
Common Crow	*Corvus brachyrhynchos*
Northern Mockingbird	*Mimus polyglottos*
American Robin	*Turdus migratorius*
Loggerhead Shrike	*Lanius ludovicianus*
European Starling	*Sturnus vulgaris*
Common Yellowthroat	*Geothlypis trichas*
House Sparrow	*Passer domesticus*
Eastern Meadowlark	*Sturnella magna*
Brown-headed Cowbird	*Molothrus ater*
Vesper Sparrow	*Pooecetes gramineus*
Lark Sparrow	*Chondestes grammacus*
Song Sparrow	*Melospiza melodia*

Amphibians and Reptiles

Tiger salamander	*Ambystoma tigrinum*
Cricket Frog	*Acris crepitans*
Woodhouse's Toad	*Bufo woodhousei*
Bullfrog	*Rana catesbeiana*
Red-eared Turtle	*Chrysemys scripta*
Ornate Box Turtle	*Terrapene ornata*

Texas Horned Lizard	*Phrynosoma cornutum*
Fence Lizard	*Sceloporus undulatus*
Ribbon Snake	*Thamnophis proximus*
Kingsnake	*Lampropeltis getulus*
Coachwhip	*Masticophisflagellum*
Copperhead	*Agkistrodon contortrix*
Western Diamondback Rattlesnake	*Crotalus atrox*

This list is not meant to be comprehensive but rather it gives a few of the more common, widespread species.

Pineywoods

The Pineywoods area contains approximately 6,000,000 hectares (15,000,000 acres) of gently rolling to hilly forested land. This is part of a much larger region of pine-hardwood forest that extends into Louisiana, Arkansas and Oklahoma. Elevation in Texas is 60 (200 feet) to 150 meters (500 feet).

This is an area of high rainfall with annual averages of 90 centimeters (35 inches) to more than 150 centimeters (60 inches). Rainfall is fairly uniformly distributed throughout the year although no single year likely will follow this pattern. Humidity and temperatures usually are high and the area is comparatively free from persistent winds.

Soils are mostly light colored to dark-gray sands or sandy loams over a clayey subsoil. Soil reaction usually is acid.

The area is interspersed with native vegetation, pasture and farm lands. Ranches are extremely varied in size and in type of operation. The principal class of livestock is cattle.

The major commercial timber species are loblolly, shortleaf, longleaf and slash pines. Many hardwoods, such as oak, hickory and maple, also are present in the overstory. Most ecologists consider the pines as a sub-climax or fire disclimax.

Subregions

Longleaf Pine Forest

Mixed Pine-Oak Forest

The Longleaf Pine Forest once dominated the east central southeastern part of the Pineywoods of Texas. A few pockets of longleaf may still be seen today, but most have been lost due to overcutting and replaced by other types of forest communities.

The Mixed Pine-Oak Forest occurs to the west and north of the longleaf pine area. Swamps are common in the southern part of the Pine-Oak Forest.

Flora

Upland

Climax	*Successional*
White Oak *(Quercus alba)*	Longleaf Pine *(Pinus palustris)*
Southern Red Oak *(Q. faicata)*	Shortleaf Pine *(P. echinata)*

Black Hickory *(Carya texana)*

Loblolly Pine *(P. taeda)*

Red Maple *(Acer rubrum)*

Sweetgum *(Liquidambar styraciflua)*

Winged Elm *(Ulmus alata)*

Post Oak *(Quercus stellata)*

Flowering Dogwood *(Cornusflorida)*

American Beautyberry *(Callicarpa americans)*

Yaupon *(Ilex vomitoria)*

American Holly *(I. opaca)*

Greenbriar species *(Smilax spp.)*

Lowland

Floodplains

Chestnut Oak *(Quercus prinus)*
Southern Magnolia *(Magnolia grandiflora)*
American Beech *(Fagus grandifolia)*
Ash species *(Fraxinus spp.)*
Sweetbay Magnolia *(Magnolia virginiana)*
Loblolly Pine *(Pinus taeda)*

River Margins

Black Willow *(Salix nigra)*
River Birch *(Betula nigra)*
Bald Cypress *(Taxodium disticum)*
Sycamore *(Platanus occidentalis)*

Fauna

Mammals

Short-tailed Shrew	*Blarina brevicauda*
River Otter	*Lutra canadensis*
Red Wolf	*Canis rufus*
Eastern Gray Squirrel	*Sciurus carolinensis*
Eastern Flying Squirrel	*Glaucomys volans*
Cotton Mouse	*Peromyscus gossypinus*
Northern Rice Rat	*Oryzomys palustris*
Florida Wood Rat	*Neotoma floridana*
Nutria	*Myocastor coypus*
Swamp Rabbit	*Sylvilagus aquaticus*

Birds

Wood Duck	*Aix sponsa*
Pileated Woodpecker	*Dryocopus pileatus*
Red-cockaded Woodpecker	*Picoides borealis*
Eastern Wood Pewee	*Contopus virens*
Wood Thrush	*Hylocichla mustelina*
Prairie Warbler	*Dendroica discolor*
Indigo Bunting	*Passerina cyanea*

Amphibians and Reptiles

Western Lesser Siren	*Siren intermedia*
Southern Dusky Salamander	*Desmognathusfuscus auriculatus*
Gray Tree Frog	*Hyla chrysosrelis*
Southern Leopard Frog	*Rana sphenocephala*

Alligator Snapping Turtle	*Macroclemys temmincki*
Eastern Box Turtle	*Terrapene carolina*
Mud Snake	*Farancia abacura*
Broad-banded Water Snake	*Nerodia fasciata*
Pygmy Rattlesnake	*Sistrurus miliarius*
Timber Rattlesnake	*Crotalus horridus*

Gulf Prairies and Marshes

The Gulf Prairies and Marshes area occupies approximately 3,850,000 hectares (9,500,000 acres) along the coast of Texas. There are two major divisions in this region, Coastal Prairie and Gulf Coast Marshlands. The Coastal Prairie is a nearly level, slowly drained plain less than 45 meters (150 feet) in elevation, dissected by streams which flow into the Gulf. The Coast Marsh is limited to narrow belts of low wet marsh immediately adjacent to the coast.

Average annual rainfall varies from less than 50 centimeters (20 inches) in the west to about 125 centimeters (50 inches) in the east. The growing season usually is more than 300 days, with warm temperatures and relatively high humidity. On the average, rainfall is fairly uniformly distributed throughout the year with slight highs in September and late spring.

Soils on the Coastal Marsh are acid sands, sandy loams and clays. The upland prairie soils tend to be heavier textured acid clays or clay loams, although there are some sandy loams. In general, soils have slowly permeable profiles and are droughty in nature.

Most of the marsh is grazed by cattle in large land holdings. Ranches and rangelands of the uplands of the Coastal Prairie are interspersed with farms. The better soils are highly productive under cultivation, as improved pastures or as native range.

The climax vegetation of the Gulf Prairie is largely grassland (tall grass prairie) or post oak savannah. However, much of the area has been invaded by trees and brush such as mesquite (*Prosopis glandulosa var. glandulosa*), oaks (*Quercus spp.*), and several acacias.

Subregions

Dunes and Barrier Islands

Estuarine Zone

Upland Prairies and Woods

These subregions are fairly self-explanatory. The dunes and barrier islands define the first, and coastal wetlands comprise the second. However, the third subregion, Upland Prairies and Woods, may be further described as a mixture of woodlands developing along alluvial valleys, and prairies on coarse sandy soils. Oak mottes are a curious feature of the Upland Subregion. Also referred to as maritime woodlands, oak mottes appear as clusters of oak trees in the midst of prairies. Actually, the individual trees in an oak motte may have begun as root sprouts from one or a few trees which took hold on the sandy prairie soils.

Prairies

Climax

Little bluestem (*Schizachyrium scoparium*)
Big bluestem (*Andropogon gerardii*)
Yellow indiangrass (*Sorghastrum avenaceum*)
Switchgrass (*Panicum virgatum*)
Brownseed paspalum (*Paspalum plicatulum*)
Plains bristlegrass (*Setaria leucopila*)
Trichloris (*Trichloris pluriflora*)

Successional (and Invaders)

Splitbeard (*Andropogon ternarius*)
Gulf Muhly (*Muhlenbergiafilipes*)
Buffalograss (*Buchloe dactyloides*)
Silver bluestem (*Bothriochloa saccharoides*)
Texas wintergrass (*Stipa leucotricha*)
Threeawn (*Aristida purpurescens*)
Smutgrass (*Sporobolus indicus*)
Vaseygrass (*Paspalum urvillei*)
Carpetgrass (*Axonopus affinis*)
Huisache (*Acacia farnesiana*)
Honey mesquite (*Prosopis glandulosa var.glandulosa*)
Baccharis (*Baccharis halimifolia*)
Chinese Tallow (*Sapium sebiferum*)

Woodlands

Post Oak	*Quercus stellata*
Live Oak	*Q. virginiana*
Honey Mesquite	*Prosopis glandulosa var. glandulosa*
Hackberry	*Celtis laevigata*

Marshes

Gulf Cordgrass	*Spartina spartinae*
Marshay Cordgrass	*S. patens*
Smooth Cordgrass	*S. alterniflora*
Saltgrass	*Distichlis spicata*
Needlegrass Rush	*Juncus rosmeriana*
Common Reed	*Phragmites communis*
Nutsedges	*Cyperus spp.*
Bullrushes	*Scirpus spp.*
Glasswort species	*Salicornia spp.*
Bushy Sea Ox-Eye	*Borrichia frutescens*

Fauna

Mammals

Greater Yellow Bat	*Lasiurus intermedius*
Red Wolf	*Canis rufus*
Northern Rice Rat	*Oryzomys palustris*

(see Pineywoods and South Texas Plains)

Birds

Brown Pelican	*Pelecanus occidentalis*
Neotropic Cormorant	*Phalacrocorax brasilianus*
Mottled Duck	*Anas fulvigula*

Greater (Attwater's) Prairie Chicken	*Tympanuchus cupido*
Tricolored Heron	*Egretta tricolor*
Roseate Spoonbill	*Ajaia ajaja*
Whooping Crane	*Grus americana*
Clapper Rail	*Rallus longirostris*
Laughing Gull	*Larus atricilla*
Royal Tern	*Sterna maximus*
Seaside Sparrow	*Ammospiza maritima*

Numerous waterfowl and shorebirds are found along coastal areas particularly during the winter season.

Oak Woods and Prairies

The Oak Woods and Prairies comprises approximately 5,000,000 hectares (12,500,000 acres). Some authorities consider this area to be part of the oak-hickory forest association or deciduous forest formation. Others prefer to class the area as a part of the true prairie association of the grassland formation. This latter view is based on the fact that the understory vegetation is typically tall grass. There also is evidence that the brush and tree densities have increased tremendously from the virgin condition.

Topography is gently rolling to hilly. Elevations are 105 to 245 meters (300 to 800 feet) above sea level. Annual rainfall is 64 to 115 centimeters (25 to 45 inches). The high rainfall month usually is May or June. Soils on the uplands are light colored, acid sandy loams or sands. Bottomland soils are light brown to dark-gray and acid, ranging in texture from sandy loams to clays.

Land use in this region is variable. Most of the area is in native or improved pastures. Although the better soils are under cultivation, there are many large ranches which predominantly raise cattle. The Western Cross Timbers comprise about 1,200,000 hectares (3,000,000 acres), the Eastern Cross Timbers comprise about 400,000 hectares (1,000,000 acres), and the Oak Woodlands about 2,600,000 hectares (6,500,000 acres).

Subregions

Western Cross Timbers

Oak Woodlands (Post Oak Savannah)

Eastern Cross Timbers

Distinctions between these subregions may be summarized as differences in the degree of development of the oak-hickory forest. The Western Cross Timbers is the least well-developed due to the lessened availability of moisture. Forests here are neither as dense nor as tall as those in the Eastern Cross Timbers. There are fewer eastern species here. Pockets of deep clay loam soils support bluestem prairies. The Eastern Cross Timbers is best described as a mix of the Western Cross Timbers and the Oak Woodlands.

The Oak Woodlands contain the best developed oak-hickory forest in Texas. Moisture is more abundant; thus, flora is more dense and lush in comparison to the

Western Cross Timbers. Prairies are more common than in the other subregions. Peat bogs are found here along the Carrizo Sand Formation. Specific relationships of plants to their environment are depicted within the Oak Woodlands along the Navasota River (Figures 1.2 and 1.3).

Several rivers cross the post oak region and due to their topographic, edaphic and drainage characteristics they contribute to plant community diversity. The Navasota River is typical and its watershed will be used to illustrate these plant communities and their environmental relationships. The lower Navasota River watershed can be divided into four habitats distinguished on the basis of similar environmental and vegetational characteristics: Upland Grassland, Upland Woodland and Savannah, Transition Forest and Bottomland Forest (Figure 1.2). Within these habitats communities can be grouped into dominance types based on the leading dominant(s). Post oak is the most widespread upland dominant, while cedar elm dominates the bottomlands. Upland understory shrubs are yaupon, farkleberry, and French mulberry, while in the bottomlands deciduous holly, hawthorne species, rusty blackhaw and others are important. The vegetation is best viewed as a continuum with individual species distribution related to topographic position, soil texture, soil depth and drainage. In the uplands, blackjack oak and juniper are locally abundant, particularly on shallow soils, while in the bottomlands water oak, green ash, slippery and American elm, willow oak, pecan, basswood, tupelo, water hickory, black willow and cottonwood are locally prominent. The latter three typically occur along well-lighted stream or river margins.

It is generally agreed that the upland post oak savannah and the river bottom were originally characterized by a more open savannah with a ground cover of mid and tall grasses. In the uplands, observation will show that most older and larger trees have an open-growth form and a relatively short stature (10–15 m) or low site index. Both of these features suggest savannah conditions as opposed to forest conditions. Under "natural" conditions a combination of soil limitations and periodic fires produced the savannah. Within the last several decades this open savannah, particularly in the uplands, has been converted into a dense woodland. This "thicketization" has occurred as a result of overgrazing, abandonment from cultivation and the removal of fire. Open, upland grasslands of the region originally occurred primarily on the clay soils which are generally unfavorable for tree growth, especially in the presence of a good grass cover and periodic fire. Mesquite is a major invader on these sites as a result of overgrazing and soil disturbance.

Upland grasslands under pristine conditions or with good grazing management support communities dominated by little bluestem, Indiangrass, Florida paspalum, switchgrass, thickspike tridens, tall drop-seed, big bluestem and others (Figure 1.3). These species have been replaced due to disturbance, by brownseed paspalum, many species of lovegrass and panic grasses, Texas wintergrass, rattail smutgrass, broomsedge and annual and perennial threeawns, as well as, many broadleaved species such as western ragweed and croton. These disturbance species, which form the seral stages of secondary succession, are generally less productive and of lower quality for large herbivores than the bluestem type climax grassland. In a general way, the major grassland successional stages are: (1) annual forbs (e.g., croton), (2) annual forbs and annual grasses, (3) annual grasses (e.g., old field threeawn), (4) annual and perennial grasses (e.g., broomsedge, silver bluestem), (5) perennial

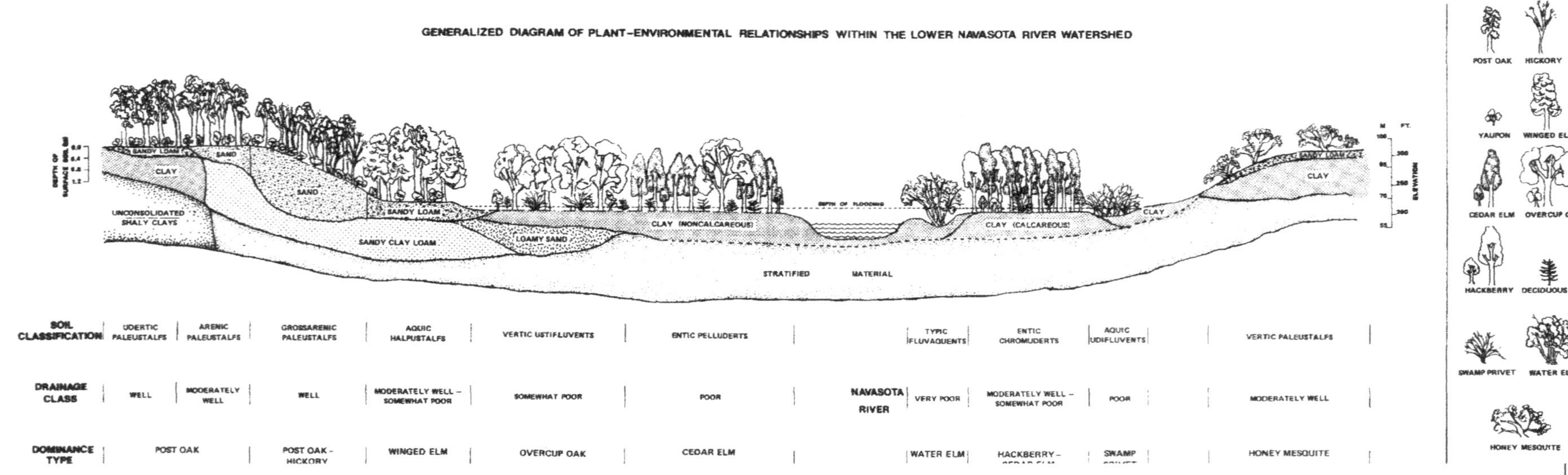

FIGURE 1.2 A generalized topographic-soil-vegetation diagram of major habitats within the Oak Woodlands (Post Oak Savannah).

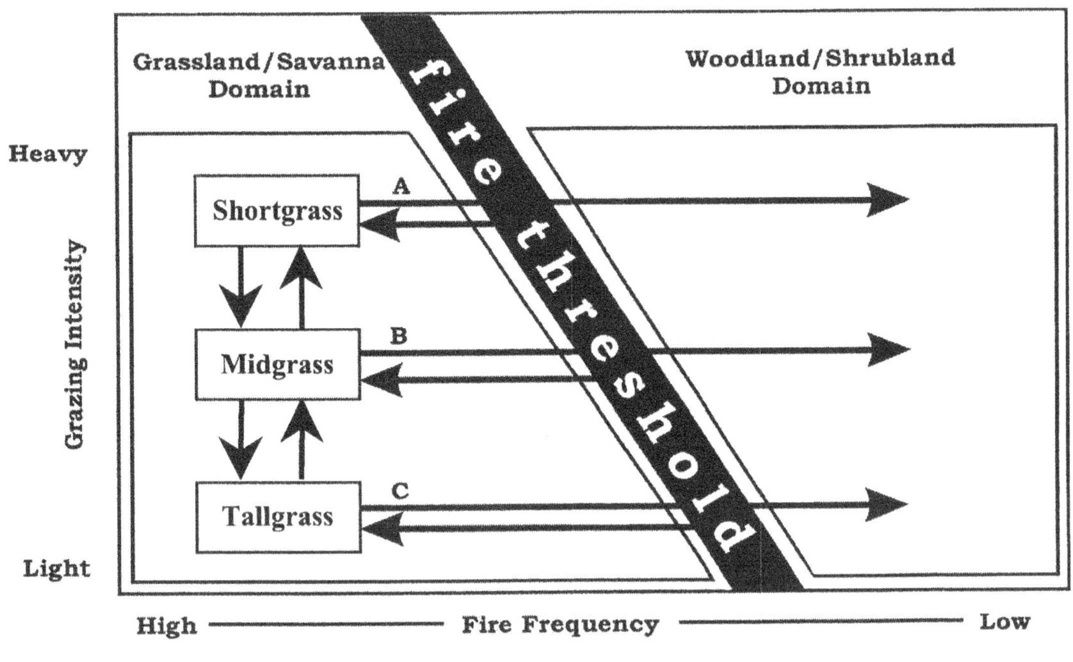

FIGURE 1.3

grasses (e.g., little bluestem), and perennial forbs (e.g., catclaw sensitive-briar). These grasslands, in the absence of fire, tend over time to be invaded by shrubs and trees similar to Post Oak Savannah.

Bottomland grassland is also generally in poor condition. The original herbaceous species consisted of Virginia and Canada wildrye, Indiangrass, switchgrass, eastern gamagrass, big bluestem, switchcane and the more productive sedges. These species have been replaced by a variety of unproductive, often unpalatable sedges and many broadleaved weedy species. The particular species present and their abundance is, of course, also influenced by soil characteristics and amount of overstory canopy cover. Generally, as canopy cover closes, herbaceous production is greatly reduced.

The post oak savannah has a variety of wildlife species (Figure 1.4). The most common large mammals are the white-tailed deer, raccoon, coyote, eastern cottontail, striped skunk, fox and gray squirrels, opossums, nutria (an introduced species) and armadillo. Less common mammals are the ringtail, red and gray fox, beaver and black-tailed jackrabbit. Small mammals include the hispid cotton rat, fulvous harvest mouse, white-footed mouse, pygmy mouse, plains pocket gopher, and Florida wood rat.

Some animals are found in many kinds of habitats while others are restricted in distribution (Figure 1.4). The white-tailed deer is found in most habitats of the post oak savannah except in intensively agricultural areas such as parts of the Brazos riverbottom. Its numbers vary with amount of food and cover and with degree of protection from harassment by man, dogs, and other disturbances. The gray squirrel is a mammal with a restricted distribution. It is found only in river bottom forest with large numbers of mature oaks, which furnish food and nest sites for the

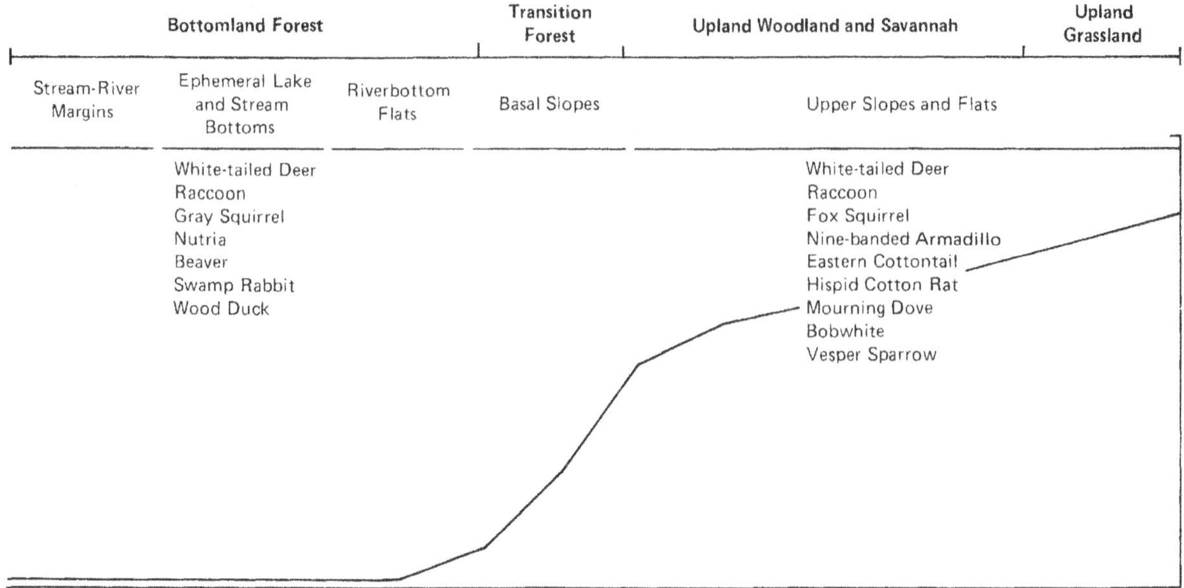

FIGURE 1.4 A generalized topographic diagram of major habitats within the Oak Woodlands (Post Oak Savannah). Common animals of bottomland and upland habitats are listed.

species. The swamp rabbit is likewise restricted to bottomland environments, while the cottontail is primarily an upland species.

Raccoons are another example of a widespread species. While they tend to prefer areas close to streams or ponds they may spend much of their time some distance from water. The beaver, on the other hand, is almost always found near water except during periods of migration.

Some species use one type of habitat at one time of the year and another type at other times. For example, some white-tailed deer in some areas may spend one season (spring) in the uplands while most of the summer may be spent in the bottomland habitats.

Less conspicuous animals such as the hispid cotton rat also relate to specific habitats. These animals are an important prey species of many carnivorous mammals and birds. The cotton rat prefers tall, dense grassland with a heavy cover of standing grass and ground litter. Grassland with little cover or heavily wooded areas have few cotton rats.

Birds are also diverse in the area. Game birds include several species of waterfowl such as the mallard, gadwall, blue-winged teal, common snipe and American coot. These are mostly winter residents while the wood duck is the most important year round resident waterfowl species. Bobwhite (quail) and mourning dove are the major upland game birds. Populations of bobwhite are controlled largely by weather conditions during the nesting and brood rearing stages and by food and cover availability. Populations of mourning doves, which are mainly migratory, fluctuate in relation to nesting success in northern states.

Birds of prey include the Red-tailed hawk, American kestrel, Red-shouldered hawk, Northern harrier and, Great-horned owl. Major scavengers are the Turkey and Black vultures and the American crow.

Songbirds are numerous and the more common ones are the Northern cardinal Scissor-tailed flycatcher (in summer), Blue jay, American robin (in winter), Carolina wren, Northern mockingbird, Loggerhead shrike, Yellow-rumped warbler (in winter), House sparrow and Eastern meadowlark. Less obvious but interesting species include the secretive Yellow-billed cuckoo and the Brown thrasher of dense brush thickets. The insect eating woodpeckers, the most abundant being the Downy woodpecker, the Northern flicker and the Yellow-bellied sapsucker, may be important control agents for many insect populations.

The European starling, Common grackle and Brown-headed cowbird are currently of importance in agricultural and urban areas of the United States. Populations of these species are very high during the winter and, at least locally, they can become serious pests.

Common amphibians and reptiles include the Gulf Coast Toad, Green Tree Frog, Leopard Frog, Ground Skink, Box Turtle, Red-eared Slider, Eastern and Western Hognose, Copperhead, Diamond-backed and Timber Rattlesnake and Water Moccasin. The latter four species, along with the Coral Snake, are important, of course, because of their venomous bites.

Insects are diverse and detailed discussion is beyond the scope of this exercise, however, a few will be mentioned. Some insects are beneficial such as the various bees and butterflies which are important pollinators. Many beetles contribute to decomposition of plants and animals and are thus important in nutrient cycling. The ladybird beetle is an important insect predator and is useful to have in gardens. Of course, some insects can be detrimental and obvious examples are the cotton boll weevil and imported fire ant. These species are economic or nuisance problems and require certain kinds of control to maintain their populations at acceptable levels. Ticks and chiggers (larvae of the Trombiculidae family) also are locally abundant and in some instances may be vectors of rocky mountain spotted fever (lone star tick) and tularemia (rabbit fever).

Blackland Prairies

The Blackland Prairies area has about 4,650,000 hectares (11,500,000 acres), including the San Antonio and Fayette Prairies. Topography is gently rolling to nearly level, well dissected with rapid surface drainage. Elevations above sea level are 90 to 245 meters (300 to 800 feet).

Average annual rainfall varies from 76 centimeters (30 inches) on the west to slightly more than 100 centimeters (40 inches) on the east. The monthly distribution pattern shows highs in May to the north to an almost uniform pattern in the southcentral region.

Blackland soils are fairly uniform dark-colored calcareous clays interspersed with some gray acid sandy loams. For the most part, this fertile area has been brought under cultivation, although some excellent native hay meadows and a few ranches remain.

Subregions

Blackland Prairies

Grand Prairie

The Blackland Prairies subregion lies on fairly deep, dark, alkaline clay soils. Tall grass types dominate.

The Grand Prairie lies on shallower soils and so does not develop into the tall grass associations. Mid-grasses are dominant.

Flora

Upland Prairie

Climax	*Successional (and Invaders)*
Little bluestem *(Schizachyrium scoparium)*	Texas wintergrass *(Stipa leucotricha)*
Yellow indiangrass *(Sorghastrum avenaceum)*	Texas grama *(Bouteloua ridigiseta)*
Big bluestem *(Andropogon gerardii)*	Silver bluestem *(Bothriochloa saccharoides)*
Switchgrass *(Panicum virgatum)*	Hairy grama *(Bouteloua hirsuta)*
Sideoats grama *(Bouteloua curtipendula)*	Buffalograss *(Buchloe dactyloides)*
Tall Dropseed *(Sporobolus asper var. asper)*	Honey Mesquite *(Prosopis glandulosa var. glandulosa)*
	Hackberry *(Celtis laevigata)*
	Post Oak *(Quercus stellata)*
	Ashe Juniper *(Juniperus ashei)*

South Texas Plains

There are about 8,000,000 hectares (20,000,000 acres) in the South Texas Plains. The topography is level to rolling and the land is dissected by streams flowing into the Gulf of Mexico. Elevations range from 305 meters (1,000 feet) to about sea level.

Average annual precipitation is 40 to 90 centimeters (16 to 35 inches), increasing from west to east. Periodic droughts are common. Average monthly rainfall is lowest during January and February and highest in May or June. After a midsummer depression, another peak is reached in September. Summer temperatures are high, with extremely high evaporation rates in the west.

Soils range from clays to sandy loams, and vary in reaction from calcareous to slightly acid. A wide range of soil profile types is responsible for great differences in soil drainage and moisture-holding capacities. Saline soils occur locally.

Although there are large acreages of good cultivated land on the South Texas Plains, most of the area is rangeland used predominantly as large cattle ranches.

This area originally supported a grassland or savannah type climax vegetation. Long continued grazing, elimination of fires, and other factors have altered the plant communities to such a degree that the region now is largely covered by dense brush. Many species of trees and shrubs have increased in the area, including mesquite, post and live oak, cacti and several acacias.

Subregions

Subtropical Zone
Brush Country
Bordas Escarpment

The Subtropical Zone is a highly modified subregion on the southern tip of Texas. The original vegetation is the result of certain climatic factors such as high rainfall. Subtropical species such as ebony and anacua dominate the relict plant communities.

A mixture of tall brush on deep soils with mesquite, spiny hackberry, acacias and short dense brush on caliche soils characterize the Brush Country.

The Bordas Escarpment is a continuous mass of short mixed brush, probably the largest single mass, on a catiche questa.

Flora

Grass Species

Sandy Loam	*Clay*
Seacoast bluestem *(Schizachyrium scoparium var. littoralis)*	Silver bluestem *(Bothriochloa saccharoides)*
Plains bristlegrass (Setaria leucopila)	Arizona cottontop *(Digitaria californica)*
Paspalum species (Paspalum spp.)	Buffalograss *(Buchloe dactyloides)*
Trichloris species *(Trichloris spp.)*	Common curlymesquite *(Hilaria belangeri)*
Chloris species *(Chloris spp.)*	Pappophorum species *(Pappophorum spp.)*
Big sandbur *(Cenchrus myosuroides)*	Alkali sacaton *(Sporobolus airoides (saline sites))*
Tanglehead *(Heteropogon contortus)*	

Woody Species

Honey Mesquite *(Prosopis glandulosa var. glandulosa)*
Post Oak *(Quercus stellata)*
Live Oak *(Quercus virginiana)*
Anacua *(Ehretia anacua* (Subtropical Zone))
Texas ebony *(Pithecellobiumflexicaule)*
Blackbrush *(Acacia rigidula)*
Guajillo *(A. berlanderi)*
Huisache *(A. farnesiana)*
Cenizo *(Leucophyliumfrutescens)*
Guayacan *(Porlieria angustifolia)*
Granjeno *(Celtis pallida)*
Anachucita *(Cordia boissieri)*
White brush *(Aloysia lycoides)*
Cactus species *(Opuntia spp.)*

Fauna

Mammals

South Texas Pocket Gopher	*Geomys personatus*
Collared Peccary	*Pecaritajacu*
Mexican Ground Squirrel	*Spermophilis mexicanus*

Birds

Harris' Hawk	*Parabuteo unicinctus*
Caracara	*Caracara cheriway*
Chachalaca	*Ortalis vetula*
Scaled Quail	*Callipepla squamata*
White-winged Dove	*Zenaida asiatica*
Ground Dove	*Columbina passerina*
Inca Dove	*Scardafella inca*
Green Kingfisher	*Chloroceryle americans*
Green Jay	*Cyanocerax yncas*
Long-billed Thrasher	*Toxostoma longirostre*
Pyrrhuloxia	*Pyrrhuloxia sinuata*
Black-bellied Whistling Duck	*Dendrocygna autumnalis*

Amphibians and Reptiles

Rio Grande Frog	*Syrrhophus cystignathoides*
Sheep Frog	*Hypopachus variolosus*
Texas Tortoise	*Gopherus berlandieri*
Rose-bellied Lizard	*Sceloporus variabilis*
Keeled Earless Lizard	*Holbrookia propinqua*
Northern Cat-eyed Snake	*Leptodeira septentrionalis*
Indigo Snake	*Drymarchon corais*

Coastal Sand Plains

The Coastal Sand Plains cover 1,036,000 hectares (2,500,000 acres) with fairly level topography. Elevations are less than 45 meters (150 feet) above sea level. Surficial and windblown sands and dune characterize this region's soils.

Precipitation varies from 90 to 115 centimeters (35 to 45 inches) per year. Vegetation is primarily grassland types but with extensive oak mottes and salt marshes including sacaton and small areas of brush. The oak scrub has become much more extensive at the expense of grassland.

Virtually all of this region has been moderately to severely grazed by domesticated cattle. Limited crop production occurs in the Sand Plains. In the past, the region has been called the "Wild Horse Prairie" because of the large herds of feral horses roaming there in the 19th Century.

Flora

Similar to the Coastal Prairies and Marshes.

Edwards Plateau

The Edwards Plateau area comprises about 9,700,000 hectares (24,000,000 acres) of "Hill Country" in West-Central Texas. On the east and south, the Balcones Escarpment forms a distinct boundary to the Edwards Plateau, but on the north and west it blends into adjacent areas. Elevations range from slightly less than 30 meters (100 feet) to more than 900 meters (3,000 feet). The surface is rough and well-drained and is dissected by several river systems.

Average annual rainfall varies from less than 38 centimeters (15 inches) in the west to more than 25 centimeters (33 inches) in the east. On the average, there are more years with below average rainfall than above. Droughts have occurred in the area frequently, but the most severe drought on record was during 1950–58. The seasonal rainfall pattern shows a typical high in May and June and again in September. This pattern shifts to a late summer high on the western edge of the Stockton Plateau.

Soils are usually shallow with a wide range of surface textures. They are underlain by limestone or caliche.

The Edwards Plateau predominantly is rangeland, with cultivation largely confined to the deeper soils and valley bottoms.

Subregions

Balcones Canyonlands

Live Oak-Mesquite Savannah

Lampasas Cut Plain

The Balcones Canyonlands subregion has the most rugged topography in the Edwards Plateau, with its steep grades and exposed geological strata. Springs abound amid the scarp woodlands of oaks and mesquite. Many plants are endemic to the area.

Live Oak-Mesquite Savannah topography is flat to rolling with oak and mesquite woods on grassland.

Grassland with scattered mesquite woods on low rolling hills underlain by limestone describes the Lampasas Cut Plain. Grasslands are found in alluvial valleys and canyon land species on slopes.

Flora

Grass Species

Little bluestem *(Schizachyrium scoparium)*
Cane bluestem *(Bothriochloa barbinodis)*
Yellow indiangrass *(Sorghastrum avenaceum)*
Texas wintergrass *(Stipa leucotricha)*
Common curlymesquite *(Hilaria belangeri)*
Tobosa *(H. mutica)*
Sideoats grama *(Bouteloua curtipendula)*
Hairy grama *(B. hirsute)*
Buffalograss *(Buchloe dactyloides)*

Wright's threeawn *(Aristida wrightii)*
Red grama *(B. trifida)*
Texas cupgrass *(Eriochloa sericea)*
Tridens species *(Tridens spp.)*

Woody Species

Live Oak *(Quercus virginiana)* elk%
Vasey Shin Oak *(Q. pungens var. pungens)*
Texas Oak *(Q. shumardii var. texana)*
Post Oak *(Q. stellata)*
Ashe Juniper *(Juniperus ashei)*
Redberry Juniper *(J. pinchoti)*
Algerita *(Berberis trifoliolata)*
Honey Mesquite *(Prosopis glandulosa var. glandulosa)*
Whitebrush *(Aloysia lycoides)*

The Llano Uplift

The Uplift area is also known as the central mineral region. It is virtually surrounded by the Edwards Plateau region. Elevation ranges from 250 to 685 meters (825 to 2,250 feet) above sea level.

Geologically the region is a large dome with rolling to hilly topography. Granite exfoliation domes, the largest of which is known as Enchanted Rock, are common. In contrast to the clays and clay loams of the Edwards Plateau, sandy soils predominate on the Llano Uplift. Rainfall averages about 75 centimeters (30 inches) with peak in May or June and September.

Oak and oak-hickory woodlands are common vegetation types, along with mesquite savannah and some grassland types that were once more widely distributed.

Subregions

Oak and Oak-Hickory Woodlands

Mesquite-Whitebrush Savannah

The Woodlands are confined to small pockets of sandy, well-watered soils. Over thousands of years, weathering of granite deposits has produced these coarse soils. Mesquite-Whitebrush Savannah predominates in the Llano Uplift region. The Savannah occurs on loamier soils underlain by caliches.

Flora

Similar to Edwards Plateau.

Rolling Plains

The Rolling Plains area is a part of the Great Plains region of the central United States. The area in Texas comprises approximately 9,700,000 hectares (24,000,000 acres) of gently rolling to moderately rough topography. It is dissected by narrow intermittent stream valleys flowing east to southeast. Elevation varies from 245 to

915 meters (800 to 3,000 feet). The eastern portion is sometimes referred to as the Reddish Prairies.

Annual rainfall ranges from about 55 centimeters (22 inches) in the west to almost 75 centimeters (30 inches) in the eastern portion. Although seasonal precipitation is highly variable, May and September normally are the high rainfall months. Typically, there is a summer dry period, with high temperatures and high evaporation rates.

Soils vary from coarse sands along outwash terraces adjacent to streams, to tight clays or red-bed clays and shales. Soil reaction is neutral to slightly calcareous.

Approximately two-thirds of this area is still in rangeland. The primary class of livestock is cattle.

Subregions

Mesquite Plains
Escarpment Breaks
Canadian Breaks

The Mesquite Plains subregion typifies the Rolling Plains Region. It is a gently rolling plain of mesquite-short grass savannah. Oak, cedar and acacia are important secondary elements of the brush portion on the savannah.

Steep slopes, cliffs, and canyons occurring just below the edge of the High Plains Caprock comprise the Escarpment Breaks subregion. The Breaks are an ecotone or transition zone between the High Plains grasslands and the mesquite savannah of the Rolling Plains. Brush species, including junipers, dominate the vegetation of this subregion.

The Canadian Breaks subregion is similar to the Escarpment Breaks, but also includes the floodplain and sandhills of the Canadian River in the northern Panhandle, bounded north and south by the edge of the Caprock. It is a mixed grass prairie with some low shrubs grading from succulents and dwarf shrubs in the east to shinnery, a savannah or groveland of scattered clusters of woody species, in the west.

Flora

Grass Species

Little bluestem *(Schizachyrium scoparium)*
Sand bluestem *(Andropogon hallii)*
Sideoats grama *(Bouteloua curtipendula)*
Hairy grama *(B. hirsute)*
Blue grama *(B. gracilis)*
Red grama *(B. trifida)*
Buffalograss *(Buchloe dactyloides)*
Common curlymesquite *(Hilaria belangeri)*
Sand dropseed *(Sporobolus crytandrus)*
Western wheatgrass *(Agropyron smithii)*
Threeawn species *(Aristida spp.)*

Woody Species

Honey mesquite *(Prosopis glandulosa var. glandulosa)*
Redberry juniper *(Juniperus pinchoti)*
Shinnery oak *(Quercus havardii)*
Sand sage *(Artemisiafilifolia)*

High Plains

The Texas High Plains area comprising about 8,090,00 hectares (20,000,000 acres), is a part of the Great Plains region. It is a relatively level high plateau separated from the Rolling Plains by the Caprock Escarpment and dissected by the Canadian River Breaks. Elevation is 915 to 1375 meters (3,000 to 4,500 feet), sloping gently toward the southeast. The surface is spotted with "playa lakes" which sometimes cover more than 16 hectares (40 acres) and contain several feet of water after heavy rains. Average annual rainfall is 38 to 54 centimeters (15 to 21 inches), with some years showing as little as 30 (12) and others more than 115 centimeters (45 inches). Extended droughts have occurred in the area, the worst in 1918–19, in the 1930's and during 1951–56. Rainfall generally is low during the winter, high in April and May, low in midsummer and high in September and October. The average frost-free period is 179 to 225 days.

Soils range in surface texture from clays on hardland sites in the north to medium textures on mixed land sites and sands on the Southern High Plains. Surface soils are generally underlain with caliche accumulations at depths of 0.6 to 1.5 meters (2 to 5 feet).

Although cultivation has increased with underground water irrigation, large areas of rangeland remain on the High Plains.

Flora

Grass Species

Blue grama *(Bouteloua gracilis)*
Sideoats grama *(B. curtipendula)*
Buffalograss *(Buchloe dactyloides)*
Little bluestem *(Schizachyrium scopariuni)*
Western wheatgrass *(Agropyron smithii)*
Sand dropseed *(Sporobolus cryptandrus)*
Galleta *(Hilaria janiesii)*

Woody Species

Honey mesquite *(Prosopis glandulosa var. glandiilosa)*
Yucca species *(Yucca spp.)*
Redberry juniper *(Juniperus pinc,hoti)*
Cactus species *(Opuntia spp.)*
Sand sage *(Artemisiafilifolia)*

Trans-Pecos

The Trans-Pecos area occupies approximately 7,700,00 hectares (19,000,000 acres) of mountains and arid valleys in the extreme western part of Texas. As the name implies, the area for the most part includes the region west of the Pecos River. It is a region of diverse habitats and vegetation, varying from desert valleys and plateaus to wooded mountain slopes. Elevation is about 760 (2,500) to more than 2590 meters (8,500 feet).

The average annual precipitation over most of the area is less than 30 centimeters (12 inches). Precipitation increases at the higher elevations, to 40 centimeters (16 inches) at Fort Davis and a reported 50 centimeters (20 inches) on Mount Locke. Although records at many locations are incomplete, there is much evidence of extreme variability. The high rainfall months usually are July and August.

Soils have developed from outwash materials from the mountains and are varied in surface texture and profile characteristics. In general, soil reaction is calcareous and some areas show accumulations of alkali as a result of poor drainage.

Cultivated areas are confined largely to the irrigated valleys. Most of the land remains in native range in large holdings.

Subregions

Sand Hills

Stockton Plateau

Salt Basin

Desert Scrub

Desert Grassland

Mountain Ranges

The Sand Hills subregion consists of shin oak and mesquite on wind-blown dunes and grasslands with some little bluestem.

Flat-topped mesas and plateaus intersected by steep-walled canyons and dry washes comprise the Stockton Plateau.

Soils with high salt concentrations and gypsum dunes on a bolson area characterize the Salt Basin. The Basin has no external drainage, hence, there is this buildup of salts in its soils. Plants which grow here have a high degree of salt tolerance.

The Desert Scrub subregion is an area of low rainfall and rapid drainage. Creosote bush flats, yucca, lechuguilla, and various small-leafed plants are common. The Desert Grassland is well described by its name. It occurs in the central part of the region as a function of altitudes and soils, which are deeper with a higher clay content.

The Mountain Ranges are characterized by higher rainfall and the development of woody vegetation such as junipers, scrub oak, live oaks, pinon pine, ponderosa pine, and Douglas fir in the Chisos, Davis and Guadalupe Mountains.

Flora

Grass Species

Chino grama *(Bouteloua breviseta)*
Red grama *(B. trifida)*
Black grama *(B. eriopoda)*
Blue grama *(B. gracilis)*

Sideoats grama *(B. curtipendula)*
Sprucetop grama *(B. chondrosioides)*
Hairy grama *(B. hirsuta)*
Tobosa *(Hilaria mutica)*
Worton threeawn *(Aristida pansa)*
Tanglehead *(Heteropogon contortus)*
Texas bluestem *(Schizachyrium cirratus)*
Muhly species *(Muhlenbergia spp.)*
Pink pappusgrass *(Pappophorum bicolor)*
Buffalograss *(Buchloe dactyloides)*
Plains bristlegrass *(Setaria leucopila)*
Alkali sacaton *(Sporobolus airoides)*
Burrograss *(Scleropogon brevifolius)*
Tridens species *(Tridens spp.)*
Finestem needlegrass *(Stipa tenuissima)*

Woody Species

Desert

Creosotebush *(Larrea divaricata)*
Tarbush *(Flourensia cernua)*
Honey mesquite *(Prosopis glandulosa var. glandulosa)*
Lechuguilla *(Agave lecheguilla)*
Sotol *(Dasylirion texanum)*
Octotillo *(Fouqueria splendens)*
Spanish dagger *(Yucca torreyi)*
Candelilia *(Euphorbia antisyphilitica)*
Pricklypear cactus *(Opuntia spp.)*
Echinocactus species *(Echinocactus spp.)*
Echinocereus species *(Echinocereus spp.)*
Redberry juniper *(Juniperus pinchoti)*
Mormon-tea *(Ephedra spp.)*
Leatherstem *(Jatropha dioica)*
Blackbruah *(Acacia rigidula)*
Sacahuista *(Nolina erumpens)*

Mountains

Oneseed juniper *(J. monosperma)*
Alligator juniper *(J. deppeana)*
Pinon pine *(Pinus cembroides)*
Gray oak *(Quercus grisea)*
Graves Oak *(Q. gravesii)*
Ponderosa pine *(P. ponderosa)*
Limber pine *(P. strobiformis)*
Quaking aspen *(Pipulus tremuloides)*

Rivers and Streams

Rio Grande cottonwood *(P. wislizenii)*
Salt cedar *(Tamarix gallica)*
Willow *(Salix gooddingii)*
Honey mesquite *(Prosopis glandulosa var. glandulosa)*

Fauna

Mammals

Desert Shrew	*Notiosorex crawfordi*
Western Canyon Bat	*Pipistrellus hesperus*
Western Spotted Skunk	*Spilogala gracilis*
Desert (kit) Fox	*Vulpes macrotis*
Cougar	*Felis concolor*
Botta Pocket Gopher	*Thamomys bottae*
Banner-tailed Kangaroo Rat	*Dipodomys spectabilis*
Merriam Kangaroo Rat	*Dipodomys merriami*
Cactus Mouse	*Peromyscus eremicus*
Davis Mountains Cottontail	*Sylvilagus robustus*
Pronghorn	*Antilocarpa americans*
Mule Deer	*Odocoileus hemionus*

Birds

Scaled Quail	*Callipepla squamata*
Band-tailed Pigeon	*Columba fasciata*
White-throated Swift	*Aeronautes saxatalis*
Broad-tailed Hummingbird	*Selasphorus platycercus*
Acorn Woodpecker	*Melanerpes formicivorous*
Cassin's Kingbird	*Tyrannus vociferans*
Mexican Jay	*Aphelocoma ultramarine*
Common Raven	*Corvus corax*
Mountain Chickadee	*Parus gambeli*
Common Bushtit	*Psaltriparus minimus*
Cactus Wren	*Campylorhynchus brunneicapillus*
Crissal Thrasher	*Toxostoma dorsale*
Black-tailed Gnatcatcher	*Polioptila melanura*
Phainopepla	*Phainopepla nitens*
Black-throated Sparrow	*Amphispiza bilineata*

Amphibians and Reptiles

Reticulated Gecko	*Coleonyx reticulatus*
Mountain Short-horned Lizard	*Phrynosoma hernandosi*
Little Striped Whiptall	*Cnemidophorus inornatus*
Trans-Pecos Blind Snake	*Leoptotyphlops humilis*
Baird's Rat Snake	*Elaphe bairdi*
Texas Lyre Snake	*Trimorphodon lambda*
Mexican Black-headed Snake	*Tantilla atriceps*

Study Questions

1. Why do so many species of plants and animals occur within the State of Texas?
2. Describe the annual amounts and seasonal distribution of precipitation across Texas.
3. Are there any endangered species of animals and plants in Texas?
4. Give examples of mammals, birds, reptiles and amphibians that are found in most parts of the State of Texas.
5. What are the major pine species that occur in the Pineywoods region?
6. Which natural region contains the remnants of the Attwater's Prairie Chicken?
7. In an ecological succession, in what part of a sere would you expect annuals to predominate? Perennials? Why?
8. Which natural region contains the greater diversity of habitats and vegetation?
9. What major factors in the environment of Texas have changed since the settlement of the State?
10. What is the importance (role) of a red-tailed hawk in a community? A pocket gopher? A woodpecker? A bee? A rattlesnake?

REFERENCES

Ajilvsgs, G. 1979. *Wildflowers of the Big Thicket: East Texas and Western Louisiana.* Tex. A&M Univ. Press, College Station. TX. 361 p.

Bartlett, R. D. and P. P. Bartlett. 1999. *A Field Guide to Texas Reptiles and Amphibians.* Gulf Publishing, Houston, TX. 180 p.

Blair, W. F. 1950. *"The Biotic Provinces of Texas." Tex. J. Sci.* 2:93–117.

Campbell, L. 1996. *Endangered and Threatened Animals of Texas: Their Life History and Management.* Texas Parks and Wildlife Press, Austin, TX. 140 p.

Carr, J. T. 1969. *The climate and physiography of Texas.* Tex. Water Devel. Bd. Rep. No. 53. 27 p.

Correll, D. S. and M. C. Johnston. 1970. *Manual of the Vascular Plants of Texas.* Tex. Res. Found., Renner, Tex. 1881 p.

Davis, W. B. and D. J. Schmidly. 1995. *The Mammals of Texas.* Tex. Parks and Wildl. Dep. Austin, TX. *** p.

Dixon, J. R. 2000. *Amphibians and Reptiles of Texas,* Second Edition. Texas A&M University Press, College Station, TX. 432 p.

Evans. D. B. 1998. *Cactuses of Big Bend National Park.* University of Texas Press, Austin, TX. 96 p.

Everitt, J. H. and D. L. Drawe. 1992. *Trees, Shrubs, and Cacti of South Texas.* Texas Tech University Press, Lubbock, TX. 216 p.

Everitt, J. H., D. L. Drawe and R. I. Lonard. 1999. *Field Guide to the Broad-leaved Herbaceous Plants of South Texas: Used by Livestock and Wildlife.* Texas Tech University Press, Lubbock, TX. 286 p.

Glenn-Lewin, D. C., R. K. Peet and T. T. Veblen (Eds). 1992. *Plant Succession: Theory and Prediction.* Chapman and Hall, New York, NY. 352 p.

Godfrey, C. L., G. S. McKee and H. Oakes. 1973. *General Soils Map of Texas.* Tex. Agr. Exp. Sta., USDA, Soil Cons. Serv., College Station, TX.

Hatch, S. L. and J. Pluhar. 1993. *Texas Range Plants.* Texas A&M University Press, College Station, TX. 344 p.

Hatch, S. L., J. L. Schuster and D. L. Drawe. 1999. *Grasses of the Texas Gulf Prairies and Marshes.* Texas A&M University Press, College Station, TX. 884 p.

Haukos, D. A. and L. M. Smith. 1997. *Common Flora of the Playa Lakes.* Texas Tech University Press, Lubbock, TX. 208 p.

Johnson, E. H. 1931. *The Natural Regions of Texas.* Univ. Tex. Bull. 3113. 148 p.

Jones, F. B. 1975. *Flora of the Texas Coastal Bend.* Welder Wildl. Found. Cont. B-6. Mission Press, Corpus Christi, TX. 262 p.

Liggio, J. and A. O. Liggio. 1999. *Wild Orchids of Texas.* University of Texas Press, Austin, TX. 240 p.

Lockwood, M. W. 2001. *Birds of the Texas Hill Country.* University of Texas Press, Austin, TX. 262 p.

Loughmiller, C. and L. Loughmiller. 1984. *Texas Wildflowers: A Field Guide.* University of Texas Press, Austin, TX. 271 p.

Lyndon B. Johnson School of Public Affairs. 1978. *Preserving Texas' Natural Heritage.* LBJ School of Public Affairs, Policy Res. Project Rep. No. 31. 34 p.

McLeod, C. A. 1971. "*The Big Thicket Forest of Texas.*" *Tex. J. Sci.* 23:221–233.

Oetking, P. F. 1959. *Geological Highway Map of Texas.* Dallas Geol. Soc., Dallas, TX.

Peterson, J. and B. R. Zimmer. 1998. *Birds of the Trans-Pecos.* University of Texas Press, Austin, TX. 208 p.

Powell, A. M. 1998. *Trees and Shrubs of the Trans-Pecos and Adjacent Areas.* University of Texas Press, Austin, TX. 464 p.

Pulich, W. P. 1988. *Birds of North Central Texas.* Texas A&M University Press, College Station, TX. 472 p.

Rappole, J. H. and G. W. Blacklock. 1994. *Birds of Texas: A Field Guide.* Texas A&M University Press, College Station, TX. 372 p.

Richardson, A. 1995. *Plants of the Rio Grande Delta.* University of Texas Press, Austin, TX. 440 p.

Richardson, A. 2002. *Wildflowers and Other Plants of Texas Beaches and Islands.* University of Texas Press, Austin, TX. 271 p.

Rose, F .L. and R. W. Stradtmann. 1990. *Wildflowers of the Llano Estacado.* Texas Tech University Press, Lubbock, TX. 111 p.

Schmidly, D. J. 1977. *The Mammals of Trans-Pecos Texas: Including Big Bend National Park and Guadalupe Mountains National Park.* Texas A&M University Press, College Station, TX. 240 p.

Schmidly, D. J. 1999. *The Bats of Texas.* Texas A&M University Press, College Station, TX. 224 p.

Sellards, E. H., W. S. Adkins and F. B. Plummer. 1966. *The Geology of Texas.* Vol. 1. Stratigraphy. Univ. Tex. Bull. No. 3232. 1007 p.

Seyffert, K. D. 2001. *Birds of the Texas Panhandle: Their Status, Distribution, and History.* Texas A&M University Press, College Station, TX. 520 p.

Simpson, B. J. 1992. *A Field Guide to Texas Trees.* Gulf Publishing Co., Houston, TX. 372 p.

Smeins, F. E. and R. B. Shaw. 1978. *Natural vegetation of Texas and adjacent areas 1675–1975: A bibliography.* Tex. Agr. Exp. Sta. Misc. Pub. 1399.

Spearing, D. 1991. *Roadside Geology of Texas.* Mountain Press Publishing Company, Missoula, MT. 432 p.

Tharp, B. C. 1939. *The Vegetation of Texas.* The Anson Jones Press, Houston, TX. 74 p.

Tveten, J. and G. Tveten. 1996. *Butterflies of Houston and Southeast Texas.* University of Texas Press, Austin, TX. 304 p.

Tveten, J. and G. Tveten. 1997. *Wildflowers of Houston and Southeast Texas.* University of Texas Press, Austin, TX. 320 p.

Vines, R. A. 1984. *Trees of Central Texas.* University of Texas Press, Austin, TX. 423 p.

Vines, R. A. 1985. *Trees of East Texas.* University of Texas Press, Austin, TX. 538 p.

Warnock, B. H. 1970. *Wildflowers of the Big Bend Country, Texas.* Sul Ross State Univ., Alpine, Tex. 157 p.

Warnock, B. H. 1974. *Wildflowers of the Guadalupe Mountains and the Sand Dune County, Texas.* Sul Ross State Univ., Alpine, Tex.

Warnock, B. H. 1977. *Wildflowers of the Davis Mountains and Marathon Basin.* Sul Ross State Univ., Alpine, Tex. 276 p,

Watson, G. 1975. *Big Thicket Plant Ecology: An Introduction.* Big Thicket Museum. Saratoga, TX.

Wauer, R. H. and C. M. Fleming. 2002. *Naturalist's Big Bend: An Introduction to the Trees and Shrubs, Wildflowers, Cacti, Mammals, Birds, Reptiles and Amphibians, Fish, and Insects.* Texas A&M University Press, College Station, TX. 208 p.

Werler, J. E. and J. R. Dixon. 2000. *Texas Snakes: Identification, Distribution, and Natural History.* University of Texas Press, Austin, TX. 544 p.

White, M. 2002. *Birds of Northeast Texas.* Texas A&M University Press, College Station, TX. 160 p.

Assignment Questions: Natural Regions of Texas

An introduction to the range of environments and biological diversity across Texas.

1. Name of ecoregion

2. Where is this ecoregion located? How large is the region?

3. What are some characteristic features of the landscape?

4. What is the dominant vegetation type (ie. grassland, forest)? Soil type?

5. Are there introduced/exotic species in the ecoregion (plants and animals)?

6. Is there extensive residential or commercial development?

7. What is the dominant land use (agriculture?)

8. Has the land use affected native plants and animals?

9. Are there any endangered or endemic species that occur in this ecoregion?

10. If there is a history of agriculture, how has agriculture affected this region?

11. Why is it important to educate the public about this particular ecoregion?

Environmental Factors

Many biotic and abiotic factors interact to control the distribution and abundance of organisms. Climatic and topo-edaphic (topography and soil) variables are among the most influential of these factors. It is the purpose of this section to provide an introduction to the methods and instruments used in the measurement and analysis of environmental variables.

It is useful initially to define some terms related to climate and soils.

Weather—the state of the atmosphere at a particular place and time with respect to temperature, moisture, currents, and pressure. The study of weather is called meteorology.

Climate—the average weather conditions of an area over a period of time, usually years. The climate is called climatology.

Macroclimate—the general climate of a broad region including the entire depth of the atmosphere.

Microclimate—the immediate environment of an organism, the layer of the atmosphere directly in contact with the organism.

Climatic Averages—the average weather conditions for the period of record of a recording station.

Climatic Normals—the average weather conditions for consecutive 30-year intervals for a recording station (e.g., 1900–1930, 1910–1940, 1920–1950, etc.).

Pedology—study of the origin and genesis of soils.

Edaphology—study of soils as a medium for plant growth and a habitat for animals.

REFERENCES

Ahrens, C. D. 2000. *Meteorology today: an introduction to weather, climate, and the environment,* 6th Edition. Pacific Grove, CA. 528 p.

Cox, G. W. 2002. *Laboratory Manual of General Ecology, 8th Edition.* McGraw-Hill, New York, NY. 320 p.

Smith, R. L. 1990. *Ecology and Field Biology.* Harper and Row Publ., New York, NY. 4th Edition. 835 p.

2.1. Measurement of Environmental Factors

It is the purpose of this chapter to provide an introduction to the methods and instruments used in measuring environmental variables related to climatic factors. Both, electronic sensors used in modern weather stations and more traditional instruments will be examined because the latter still have wide utilities and are better suited for illustrating the mechanisms of the instruments.

Radiant Energy

The sun is the source of the earth's radiant energy. Solar radiation of direct significance to organisms falls in the 0.4–3.0 micron (10^{-6}m) wavelengths of the electromagnetic spectrum. This range can be divided into the 0.4–0.7 micron and the 0.7 to 3.0 micron portions that represent the visible (light) and infrared (thermal) wavelengths, respectively. Characteristics and measurement of these two portions will be discussed in the sections, Light and Temperature.

All wavelengths of energy can be transformed to progressively longer wavelengths according to the First Law of Thermodynamics. Thus, all energy from the sun can be transformed to infrared (thermal) energy and the resultant thermal energy measured to provide a measure of total solar energy. Many devices of varying sophistication have been developed to make this measurement. A commonly used instrument for field measurement of above-the-canopy radiation is the **pyrheliometer**. It is designed to provide a continuous record of the intensity of total radiation energy from the sun and sky. This includes indirect radiation scattered by the atmosphere and reflected from clouds. Radiation passes through a glass hemisphere and strikes two bimetallic strips, one white and the other blackened with a dull finish. The white strip reflects the radiation and remains at the temperature of the surrounding air. The black strip, on the other hand, is warmed by incident radiation and tends to expand and curve. The difference in curvature of the two strips is proportional to their temperature difference and therefore to the intensity of the incident radiation. Changes in the curvature difference amplified by linkage moves the pen arm across the chart of the drum and changes in solar radiation are displayed as pen markings on the chart. The glass hemisphere covering the bimetallic sensors is transparent to wavelengths of electromagnetic radiation energy between 0.36 microns and 2.5 microns.

Electronic solar radiation sensors, such as **pyranometers**, are often used in automated weather stations. Pyranometers based on the thermopile technology use a blackened thermopile (a series of thermoelements) as sensor and assure very broad spectral sensitivity, almost ideal spectral response, and long-term stability. Pyranometers is often used with wavelength blocking filters in order to measure radiant power distribution in various bands.

A great variety of units are utilized to express energy relationships. However, a familiar unit, the calorie, will be used here. One calorie is equivalent to the energy required to raise the temperature of 1 gram of water from 14.5 to 15.5°C. To express energy flux the unit cal/cm^2/min is used.

Light

A useful instrument for measurement of the visible (light) portion of the electromagnetic spectrum is the photoelectric **photometer.** Photometers are calibrated in units of illumination. In the metric system the units are lumens/m^2 (lux) or in the English system the units are lumens/ft^2 (ft-candles). The relation between the two systems is 1,000 lux equals 93 ft-candles. As a general relationship 1 cal/cm^2/min of solar energy provides 6,700 ft-candles of illumination on a clear day with the sun at zenith.

Other approaches to light measurement have been developed but will not be considered here. In the absence of refined equipment, good approximations of light intensity may be obtained with photographic light meters even though they are not calibrated in ft-candles.

Temperature

There are several kinds of instruments, such as various types of thermometers, for temperature determination. For many purposes, the ordinary **mercury bulb thermometer** is a suitable instrument. To measure air temperatures, the thermometer should be kept in the shade, exposed to the wind and away from the influence of one's body. Special metal encased thermometers are available for determining soil temperatures.

The **maximum-minimum thermometer** indicates the temperature extremes reached over a period of time (e.g., 24 hours). It consists of a U-tube with a glass bulb at one end that is filled with a liquid that expands or contracts with increase or decrease of temperature, and an expansion chamber at the other end. The connecting U-tube contains the mercury column and part of the expanding liquid. There are two markers that are pushed to the temperature extremes by the two sides of the mercury column and are held in position to indicate the maximum/minimum temperatures. To reset, use a magnet to draw the markers or press the release button, until the markers come to rest on top of the mercury columns.

Continuous temperature records are obtainable with **thermographs.** These usually consist of a bimetallic or liquid-in-metal expansion element attached by levers to a pen, which records on a graduated chart revolving on a drum. Types with two and three sensitive elements and pens, attached by flexible cables, permit simultaneous recording of temperatures at different levels of soil, air, or water.

Electronic thermometers with **thermistors** as the sensing element, are widely used today (e.g., tele-thermometer). A thermistor consists of a semi-conductor that has a unique property that its resistance to electrical currents is a function of its temperature. Thus the thermistor's temperature can be measured by measuring the resistance of the thermistor. These devices have the advantage that they give rapid responses to change in temperature, and can be used with a data logger to provide near continuous temperature records. Thermistor probes can be obtained in a variety of forms. Most useful for soils work is a probe completely enclosed in steel. These probes have the advantage that they can be buried in the soil for periods of a year or more, and readings may be taken as often as desired without disturbing the soil.

Temperature is recorded as degrees Fahrenheit or degrees Celsius. Celsius temperatures may be converted to Fahrenheit units by means of the formula:

$$°F = \frac{°C \times 9}{5} + 32$$

and Fahrenheit units to Celsius by means of the formula:

$$°C = (°F - 32) \times \frac{5}{9}$$

Humidity

Relative humidity of the atmosphere may be measured by a wet and dry bulb **psychrometer** that consists of a pair of matched thermometers, one of which is provided with a wick around the bulb. The thermometers are mounted in a device that provides for the circulation of air over their bulbs. In a sling psychrometer this is merely a rod that may be revolved rapidly about a pivot. The cog psychrometer is a modified eggbeater, which revolves the thermometer bulbs in a single plane. The most accurate one is the hand or motor aspirated psychrometer that draws a stream of air over the stationary bulbs. In all types, the wick covering one of the thermometer bulbs is wetted with distilled water (the 'wet bulb'). The instrument is then operated by rotation or aspiration until the wet bulb temperature has reached a minimum value, as determined by frequent inspection. At this point, both dry bulb and wet bulb temperatures are recorded. Relative humidity (RH) is then calculated from the dry bulb temperature and the difference between the wet and dry bulb temperatures, by means of a conversion sliding-ruler.

A **hygrograph** is a continuously recording instrument in which an arm marks on a rotating drum the expansion and contraction of strands of human hairs that respond to changes in relative humidity. The **hygrothermograph** is a combination temperature and relative humidity recorder. Dual sensing elements ink their traces on a single chart. The temperature pen is actuated by either a liquid-filled bourdon tube, or a bimetallic element. The humidity pen's movement is caused by the increase or decrease of humidity acting on a human hair element causing it to expand or contract. The instrument should be installed in a shelter that permits free airflow.

Electronic humidity sensors, such as the capacitive thin-film humidity sensors, are often used in automated weather stations. The thin polymer film either absorbs or exudes water vapor as the relative humidity of the ambient air rises or drops. The dielectric properties of the polymer film depend on the amount of water contained in it: as the relative humidity changes, the dielectric properties of the film change and so the capacitance of the sensor changes. The electronics of the instrument measure the capacitance of the sensor and convert it into a humidity reading.

Precipitation

Precipitation can be measured with a **standard rain gauge** (U.S. Weather Bureau), which is a cylinder 8 inches in diameter and 20 inches high. A funnel built into the upper end permits the water it catches to run into an inner cylinder with exactly one-tenth the cross-sectional area. The ratio of the outer to the inner cylinder being 10:1 makes accurate readings possible to 0.01 inch. Exceptionally heavy rains may overflow the tube, and the water in the large cylinder must then be poured over into the emptied tube for measurement.

Several types of recording gages have been devised and they often either: (1) register and record small increments of precipitation, or (2) weigh and record accumulative precipitation as it falls. The most often used is the **tipping bucket rain gage.** It consists of a funnel and a rocker mechanism with two little "buckets" on it, located underneath the funnel. The funnel catches and funnels rain into one of the buckets on the rocker mechanism at a steady rate. Once a given amount of rain has been collected, usually 0.01 inches (0.25 mm), the rocker "tips" over, emptying one of the buckets and moving the other one underneath the funnel. Each tip is marked by a switch closure that is recorded by a datalogger. Because the water drains through the base after measurement, the tipping bucket rain gage will not overflow as the standard rain gage and can be deployed for long periods of time. Some tipping bucket rain gages are equipped with a heating system for collecting precipitation in freezing weather. **Weighing bucket rain gage** is another type of recording gage. It consists of an antifreeze-filled collection bucket mounted on top of a scale. As precipitation of any form falls into the collecting bucket, the bucket becomes heavier. The device is calibrated so that the weight of rainfall is recorded directly in terms of rainfall in inches or millimeters. The accumulative precipitation is either recorded continuously by marking on a chart on a rotating drum or recorded electronically at fixed time intervals.

Wind

Wind speed is measured using the totalizing **anemometer,** a three-cup device that records wind passage in distance. This instrument does not measure the instantaneous wind speed, but determines the average wind speed during a given time interval by dividing units of wind passage by elapsed time. A modern wind speed sensor often contains a 3-cup anemometer and a sealed magnetic reed switch. Rotation of the cupwheel produces a high frequency pulse that is directly proportional to wind speed. The frequency of the pulse is measured by a datalogger and converted to wind speed units such as miles per hour (mph), meters per second (mps), or knots. Another kind of wind speed sensor uses a four-blade propeller. Rotation of the propeller produces an AV sine wave and the frequency is directly proportional to the wind speed.

For instantaneous wind speed measurement, a hand-held **wind meter** may be used. The electronic pocket wind meters actually use anemometer-based wind speed sensors. They can give near instantaneous wind speed readings by providing average wind speed for very short time intervals.

Wind direction sensor usually consists of a wind vane with a potentiometer. When a precision excitation voltage from a datalogger is applied to the potentiometer element, the output signal is a voltage that is directly proportional to the azimuth of the wind direction and recorded by a datalogger.

Evaporation

Evaporation is measured using **evaporimeters** that measures the rate of evaporation of water into the atmosphere. Evaporimeters are of two types, those that measure the evaporation rate from a free water surface and those that measure it from a continuously wet porous surface.

One of the most often used evaporimeters in the first type is the **Class-A Evaporation Pan.** It consists of a 10″ deep and 47.5″ in diameter stainless steel pan installed on a wooden platform set on the ground in a grassy location. The amount of water evaporated is determined based on periodic measurement of the water level using a water-level gauge, with corrections made for precipitation during the interval between readings. Since the amount of evaporation is a function of temperature, humidity, wind, and other conditions, in order to relate the evaporation to current or expected conditions, the water temperature and wind speed are normally recorded along with the evaporation.

The **Livingston atmometer** is an example of an evaporimeter of the second type. It consists of a porous clay sphere or cup connected to a reservoir by means of a tube. Water evaporating from the clay surface is constantly replaced from within and drawn from the reservoir through the tube. The reservoir is marked near the top and filled to this mark by lifting the stopper. Subsequent fillings made at regular intervals indicate water lost to the air by evaporation over the period of time involved. The simplicity of this device has been in its favor, and it has other advantages. All atmometer bulbs are standardized to permit direct comparison of results obtained with every instrument wherever it is used. The spherical form of the atmometer bulb gives it the advantage of exposing half its surface to the sun regardless of the sun's position. Black bulbs can be used in combination with white and the increased evaporation resulting from their greater heat absorption may be used as a measure of relative solar radiation in different habitats.

Study Questions

1. What portion of the electromagnetic spectrum includes visible wavelengths? Thermal wavelengths?
2. Define a calorie.
3. What is a photometer?
4. What is the advantage of a thermistor compared to a mercury bulb thermometer?
5. What is a psychrometer?
6. Why does the wet bulb thermometer of a psychrometer usually have a lower temperature than the dry bulb?
7. What instrument combines a continuous measurement and record of temperature and humidity?
8. A black and white bulb atmometer are set up in the same experiment. Which one would lose the greatest amount of water?

2.2. Moisture as an Environmental Factor

Water is an essential substance for life and is also a major factor of an organism's environment. For terrestrial systems, moisture occurs in three major "sinks": (1) atmosphere (water vapor and liquid water); (2) soil (soil moisture); and (3) lakes, rivers, etc. (ponded water). Water vapor in the atmosphere may condense to form water droplets and produce precipitation. Precipitation that falls to the earth's surface may be temporarily stored in the soil, in lakes and rivers or in plants and animals, but eventually it is either evaporated or transpired back to the atmosphere to complete the hydrologic cycle.

Atmospheric moisture may occur as either a vapor or a liquid. The quantity and seasonal distribution of this moisture is of critical importance to the distribution and abundance of organisms. This chapter deals primarily with atmospheric moisture.

Water Vapor

Water vapor is the invisible water in the air. The quantity of water vapor in the atmosphere is generally expressed in two ways: relative humidity and vapor pressure deficit. Relative humidity is the ratio of the actual number of grams of water in one cubic meter of air to the maximum number of grams which one cubic meter can hold at a given temperature. Relative humidity is usually expressed as a percent. Vapor pressure deficit is the difference between the partial pressure of water vapor in the atmosphere at a given time and the potential partial pressure of water vapor if the atmosphere were saturated. Vapor pressure deficit is usually expressed in millimeters of mercury (mm Hg).

Relative humidity is valuable for comparing the same location from time to time or for comparing two adjacent habitats (for example, an open grassland with an adjacent forest), but it is not a useful measure for comparing two geographically isolated areas at the same time. Vapor pressure deficit is a direct measure of the evaporative force of the atmosphere (the greater the deficit the greater the potential for evaporation) and thus, it is more useful for comparing two different areas than is relative humidity.

Relative Humidity

Absolute humidity is the amount of water vapor per unit volume of air, or actual water content of the air, and usually expressed as g/m^3. Saturation humidity is the maximum amount of water the air can hold at a given temperature and is determined experimentally and is available from tables (Handbook of Chemistry and Physics). Relative humidity is the ratio of absolute and saturation humidity.

$$\text{Relative humidity (RH)} = \frac{\text{absolute humidity (AH)}}{\text{saturation humidity (SH)}} \times 100$$

For example, an air mass has a temperature of 32°C, an absolute humidity of 0.60 g/m^3, and a saturation humidity of 1.0 g/m^3:

$$\text{RH} = \frac{0.60}{1.00} \times 100 = 60\% \quad \text{at } 32°\text{C}$$

Since relative humidity is temperature dependent, it varies with the temperature of the air mass as well the water content of the air mass. If the temperature in the above example is lowered to 27°C, the saturation humidity changes to 0.70 g/m^3 and the relative humidity becomes:

$$RH = \frac{0.60}{0.70} \times 100 = 85\% \text{ at } 27°C$$

TABLE 2.1 Relative Humidity (%) Reference Values.

Dry-Bulb Temp., °C	Dry-Bulb Temperature Minus Wet-Bulb Temperature °C														
	1	2	3	4	5	6	7	8	9	10	12	14	16	18	20
2	84	68	52	37	22	8									
4	85	71	57	43	29	16	3								
6	86	73	60	48	35	24	11								
8	87	75	63	51	40	29	19	8							
10	88	77	66	55	44	34	24	15	6						
12	89	78	68	58	48	39	29	21	12						
14	90	79	70	60	51	42	34	26	18	10					
16	90	81	71	63	54	46	38	30	23	15					
18	91	82	73	65	57	49	41	34	27	20	7				
20	91	83	74	66	59	51	44	37	31	24	12				
22	92	83	76	68	61	54	47	40	34	28	17	6			
24	92	84	77	69	62	56	49	43	37	31	20	10			
26	92	85	78	71	64	58	51	46	40	34	24	14	5		
28	93	85	78	72	65	59	53	48	42	37	27	18	9		
30	93	86	79	73	67	61	55	50	44	39	30	21	13	5	
32	93	86	80	74	68	62	57	51	46	41	32	24	16	9	
34	93	87	81	75	69	63	58	53	48	43	35	26	19	12	5
36	94	87	81	75	70	64	59	54	50	45	37	29	21	15	8
38	94	88	82	76	71	66	61	56	51	47	39	31	24	17	11
40	94	88	82	77	72	67	62	57	53	48	40	33	26	20	14

Vapor Pressure Deficit

Saturation vapor pressure and actual vapor pressure are used to determine vapor pressure deficit (VPD). Saturation vapor pressure (SVP) is the pressure of a column of water vapor when the air is saturated, measured in units of pressure such as inches or millimeters of mercury. This value can be found in standard tables. Actual vapor pressure (AVP) is the actual partial pressure of water vapor in the atmosphere. This value is determined based on the relative humidity of an air mass and the saturation vapor pressure ($AVP = SVP \times RH$).

Vapor pressure deficit (VPD) = saturation vapor pressure (SVP) – actual vapor pressure (AVP)

For example, an air mass has an RH of 100%, a temperature of 20°C and the SVP = 17.54 mm Hg:

$$VPD = 17.54 - (17.54 \times 1.00) = 0 \text{ mm Hg}$$

Now lower RH to 70%

$$\begin{aligned} VPD &= 17.54 - (17.54 \times 0.70) \\ &= 17.54 - 5.26 \\ &= 2.28 \text{ mm Hg} \end{aligned}$$

Vapor pressure deficits can be determined from tables developed for this purpose.

Liquid Water and Precipitation

Liquid water in the atmosphere occurs in many forms: dew, fog, smog, clouds, and precipitation (rain, snow, sleet, and hail).

Dew Point

Dew point is the temperature below which condensation is expected to occur in a given air mass. It equals 100% relative humidity or saturation vapor pressure. For example, if an air mass has these characteristics: an AH of 0.40 g/m^3, a SH of 0.65 g/m and a temperature of 26°C:

$$RH = \frac{0.40}{0.65} \times 100 = 62\% \quad \text{at } 26°C$$

Now cool the air mass to 17°C and the saturation humidity becomes 0.40 g/m^3, then

$$RH = \frac{0.40}{0.45} \times 100 = 100\% \quad \text{at } 17°C$$

Thus, dew point is 17°C or the temperature at which water vapor can turn to liquid in this particular air mass (generally also need condensation nuclei, such as salt particles, dust, etc., for the water molecules to coalesce around).

Mechanism of Precipitation

1. Requires a rising air mass.
2. The rising air mass expands as it rises due to decreased pressure.
3. As the air mass expands it cools since the molecules are farther apart (kinetic energy is reduced).

   The amount of temperature change of an air mass when it rises depends on whether condensation is occurring. Before condensation, the temperature of an air mass changes 10°C per 1,000 m rise in elevation (dry adiabatic rate). During condensation, however, the temperature of an air mass changes 6°C per 1,000 m rise in elevation (wet adiabatic rate). This is because heat of condensation (580 cal/gm of H_2O) is being added to the air mass and thus reduces the rate of temperature decrease. Adiabatic refers to temperature change without a gain or loss of heat.
4. Assuming the presence of water vapor the cooling continues until Dew Point is reached which may then result in condensation and precipitation if condensation nuclei are present and the droplets become large enough to fall under the force of gravity.

Causes of Precipitation

1. *Cyclonic or Frontal Systems*

   A large warm air mass of varying size, usually hundreds of square miles in size, rises because warm air is lighter than cold air and as the warm air rises adjacent cold air masses move in to replace it and continues to push the warm air upward—this usually results in long-lived regional type precipitation patterns. These systems are generally related to the pressure systems that exist over a continental area. Pressure is reported in inches or millimeters of mercury: 1 inch equals 34 millibars pressure. Thus, 760 mm or 30 in of mercury would be 1,020 millibars of pressure.

2. *Convectional Systems or Thermals*

   A local land surface, a few square miles or a portion of a state, is warmed more than adjacent areas and the air next to it warms and rises. This results in local erratic thunderstorm type precipitation that is short-lived. Several of these may coalesce to form a large storm.

3. *Orographic Systems*

   Moving air masses are forced over mountains or other topographic surfaces and as they rise, precipitation may occur. Below is an example of moisture and temperature change in an air mass as it goes through the precipitation mechanism.

An air mass is located at 1,000 m above sea level and has a temperature of 20°C. It rises to 3,000 m and dew point is reached. The temperature at 3,000 m is:

$$20°C - [(3{,}000\ m - 1000\ m) \times 10°C / 1{,}000\ m] = 0°C$$

The air mass continued to rise to 10,000 m with condensation occurring all the way

$$0°C - [(10{,}000\text{ m} - 3{,}000\text{ m}) \times 6°C/1{,}000\text{ m}] = -42°C$$

The air mass now runs out of water and descends from 10,000 m down to 1,000 m

$$-42°C + [(10{,}000\text{ m} - 1000\text{ m}) \times 10°C/1{,}000\text{ m}] = 48°C$$

The result is that the air mass has now not only lost most of its moisture but also has warmed up by 28°C (48 – 20) and correspondingly its evaporative power has increased.

If the above examples were to occur in relation to a mountain, the result would produce a "rain-shadow" with the windward side of the mountain relatively wet and cool and the leeward side dry and hot.

Precipitation that reaches the land surface may infiltrate into the soil, may percolate through the soil to the groundwater, may be stored temporarily in the soil or may run off the soil surface. Runoff water may be ponded in lakes and ponds or may move to streams and rivers. Physical, chemical, and biological aspects of lake and stream water are discussed in the aquatic ecosystem studies section.

Study Questions

1. What factors influence the relative humidity of an area?
2. Why is vapor pressure deficit considered to be a better way to compare two areas than is relative humidity?
3. What is the dry adiabatic rate?
4. Why is the wet adiabatic rate lower than the dry rate?
5. What are the three mechanisms that produce precipitation?
6. Explain why the leeward side of a mountain range is usually drier and hotter than the windward side.

Assignment Questions: Environmental Factors

Primary axes of terrestrial and environmental variation and their measurement

Environmental Factors

Light Sensors

1. What is the functional mechanism used in the Extech light meter? What are the units of light flux and what do they mean?

2. Name two types of light sensors commonly used in ecological research. Why might ecologists be interested in one versus the other?

Temperature and Moisture

1. What types of functional mechanisms are found in the hygrothermograph?

2. What two general mechanisms are used in modern humidity sensors?

3. What relationship is exhibited between temperature and relative humidity? Why?

4. Briefly explain the mechanism of precipitation. What are the three primary causes of precipitation?

(FIELD)

Wind Meter

1. Determine the current wind speed in the field based on a hand-held meter:

2. Does it record data on a continuous or instantaneous basis?

3. Does your measurement differ from that of your classmate made at the same time? Why or why not?

Sling Psychrometer

1. What kind of water is used to wet the wick covering? Why?

2. Using Table 2.1, determine the relative humidity

   (RH) in the field:

wet bulb temperature ______°C
dry bulb temperature ______°C
RH= ______%

Digital Thermometer / Relative Humidity Sensor

1. What types of functional mechanisms are found in this instrument?

2. Determine the current relative humidity in the field using the digital sensor:____________. How does this compare to the measurement from the sling psychrometer?

Why?

STUDY QUESTIONS

1. Field instrument designed to provide a continuous record of above-canopy total radiation intensity.

2. Wavelengths of the electromagnetic spectrum associated with a) light and b) thermal energy.

3. Distinguish between weather and climate.

4. Distinguish between macroclimate and microclimate. Distinguish between climatic averages and climatic normals. Distinguish between pedology and edaphology. Define the term calorie.

5. Instrument used to measure the visible portion of the electromagnetic spectrum.

6. Instrument with bi-metallic or liquid-in-metal expansion element for measuring continuous temperature records.

7. Electronic thermometer with thermistors that gives rapid response to temperature changes.

8. Instrument used to measure temperature extremes over a period of time.

9. Instrument using paired thermometers to measure instantaneous relative humidity.

10. Continuous recording instrument using expansion and contraction of human hair to measure changes in relative humidity.

11. Continuous recording instrument that measures both temperature and relative humidity.

12. Instrument that transmits and indicates wind passage.

13. Hand held instrument for measuring instantaneous wind speed.

14. Instrument that seeks to integrate the effects of humidity, radiant energy, and wind to water loss in a manner similar to that of plants.

15. Convert °F to °C.

16. List the three major terrestrial moisture sinks.

17. The measures of water vapor that are valuable for comparing a) the same location from time to time or two adjacent habitats and b) two different areas.

18. Does relative humidity increase or decrease as temperature increases?

19. A relative humidity calculation (equation provided).

20. A vapor pressure deficit calculation (equation provided).

21. A temperature equal to 100% relative humidity or saturation vapor pressure below which condensation is expected for a given air mass.

3

Adaptations

Adaptation may be defined as the conformity between organisms and their environment (habitat). Adaptations are genetically fixed traits that are the result of the process of **natural selection.** Since adaptations result from an ongoing process, the observed traits that organisms exhibit are the result of past events and selective forces, as well as those same, and perhaps some new, forces operating on the contemporary scene. **Genetic variation** within a species is what allows for continued selection and adaptation to occur. Those individuals that leave the most descendents are considered the most **fit** for a given environment relative to the number of descendents left by other, less fit individuals. Thus, natural selection and the resultant adaptations are never constant but potentially ever-changing. The degree and rate of change, and whether those changes are continuous or punctuated in time, is generally controlled by the degree and rate of change of the environment, both abiotic and biotic. Organisms must respond to the **conditions** in their environment, that is, non-consumed aspects of the environment such as temperature, pH, etc., and to **resources**, that is, consumed aspects of their environment, such as oxygen, water, nutrients, etc.

Adaptations may be expressed as **morphological/anatomical, physiological, life history or behavioral traits,** that enable organisms to survive, grow and reproduce in a given environment. In many habitats a particular environmental factor, such as moisture content, often appears to be the prevailing selective force; in others it may be a biological interaction such as predation. Thus, adaptations are often viewed in relation to an individual environmental factor or biotic interaction. While this is convenient, and may be appropriate, it is essential to remember that organisms must deal with and respond to a suite of factors in their environment and to place too much emphasis on single factors may lead to misinterpretations of how organisms conform to their environment. Also, morphological and anatomical features of organisms are usually more prominent and obvious features of an organism, and hence are most often used to characterize adaptation. Less obvious and more difficult to measure and observe physiological, behavioral and life history traits may often be more important to survival in a given environment. Fortunately, there is often a reasonable known or implied relationship between **form (morphology)** and **function (physiology).** Hence classifications of organisms based on their outward morphological features often implies associated physiological adaptations.

It is important to keep in mind that many individuals of species exhibit a great deal of **phenotypic plasticity** which is non-evolutionary, environmentally induced variation (e.g., see polymorphic leaves in chapter 2.1). This expression of phenotypic plasticity is itself an important feature of organisms that allows them to persist particularly in environments that fluctuate greatly during the establishment, growth and development phases, which allows the organism to adjust to the variable conditions, whereas species without such flexibility would likely perish. Also, organisms can acclimatize to varying environments. **Acclimation** is a special case of phenotypic variation which is a plastic, temporary change in an organism caused by an environment to which it has been exposed in the past. For example, creosotebuth (*Larrea divaricata*) shifts its temperature optimum for photosynthesis from about 40°C in the summer to about 20°C during the winter. This allows year round growth in this species.

The classification of, and adaptations of, organisms to several important environmental parameters are discussed in the following sections. While other factors may be of equal importance, knowledge of these factors and adaptation to them go a long way in describing and understanding the distribution and abundance of species in nature.

3.1. Plant Adaptations

Moisture

Water is an essential component of living things and as well it forms a major part of the environment within which organisms live. No plant species can survive in both very wet and very dry habitats and each species tends to flourish within a limited range of moisture conditions; some have broad ecological tolerances to moisture (**euryhydric**), while others have narrow tolerances (**stenohydric**). This, of course, is true of a species' relationship to any environmental gradient. Most plants exhibit a suite of adaptations that permit a general classification into broad categories of plants based on moisture regime of the habitat.

Plants that live wholly or partly submerged in water or in very wet places are known as **hydrophytes.** Plants of lakes, ponds, streams, and other bodies of water, both fresh and salt, belong to this group.

Most plants live in an environment that is characterized by a medium water supply. The plants of forests and prairies, as well as, many crop plants usually belong to this group. Plants of habitats that usually show neither an excess nor a deficiency of water are known as **mesophytes.**

Plants that grow where evaporation stress is high and the water supply is low, show characteristic adaptations to a decrease in water content. They are termed **xerophytes.** Certain species of deserts, alpine peaks, sand hills, and dry prairies belong to this group. As among hydrophytes, plants of this group include species from many families which are unrelated phylogenetically, although some, as a result of the impress of the environment, have come to resemble each other more or less closely in vegetative characters, as exemplified by various cacti (Cactaceae) in arid portions of America and certain spurges (Euphorbiaceae) in African deserts.

Hydrophytes, xerophytes, and mesophytes are readily distinguishable as groups related to moisture regime, since each has a more or less definite habitat and characteristic appearance, however, all gradations of form and adaptations are found due to the existence of intermediate habitats.

Xerophytes

With respect to their relation to water, xerophytes may be arranged into two groups. Most adaptations of both groups are, of course, related to reduction of water loss, increase in water uptake efficiency and reduction in temperature load on the plant.

1. Drought Evaders. Annual herbaceous plants evade the more extreme desert conditions by completing their life cycle during the short rainy season and passing the remainder of the year as fruits or seeds lying dormant in the soil. Dry seeds are often able to survive high temperatures and dehydration without losing their ability to germinate later when conditions become favorable. In addition, the whole life cycle is compressed into a few weeks, and hence such plants are sometimes termed ephemerals.

Delay in seed germination is an adaptation that is achieved by various physiological mechanisms of which the best understood is the presence of chemicals in the seed coat which act as germination inhibitors. Germination can only take place after sufficient rain has fallen to leach out the inhibiting chemical.

The possession of two or more types of seeds, which differ in their germination responses also assist in preventing plants from germinating all at the same time. In saltbushes of the genus *Atriplex,* for example, various kinds of seed are found. Some seeds are brown and germinate throughout the year after shedding; others are black and do not usually germinate until two or more years have passed.

2. Drought Resistors. Some xerophytes are characterized by the ability to survive long periods of drought and dehydration of their tissues without suffering much injury. They show extensive resistance to wilting. Water loss is resisted in a number of ways and different protective devices are often combined in the same plant. Transpiration through the cuticle may be lowered by the deposition of lipids or fatty substances and resins on the leaf surface. Other morphological characters which may be adaptations to drought include reduction in cell size, presence of dense, hairy coverings, sunken stomata, increase in palisade tissue, a diminution in size of leaves, development of sclerenchymatous tissue to give increased mechanical support, leaf-folding or rolling and leaf-shedding.

In addition to exhibiting a much reduced leaf surface and cuticular transpiration, plants such as the succulent Cactaceae and Euphorbiaceae must depend for their extreme drought resistance upon their well-developed water-storing tissues and low surface-to-volume ratio, for they are always thick and fleshy. Fluted stems, such as those of the saguaro cactus, expand and contract like an accordion as moist and dry periods alternate.

To resisting water-loss, many desert plants have evolved extensive root systems that serve to increase water-uptake. In succulents, these roots spread laterally but are seldom more than 3 or 4 centimeters from the soil surface. Thus, they exploit to the maximum every shower of rain, even if the water does not penetrate far into the soil.

The majority of perennial woody desert plants have extraordinarily deep roots. Thus, mesquite has been found with roots to a depth of over 30 meters. The long taproots may reach a water table many meters below the surface while a superficial rooting system exploits the moisture from showers and thunderstorms.

Many perennial desert plants exist beneath the ground in a dormant condition during the dry months and produce leaves, stems and flowers after rain has fallen. The perennial part of the plant takes the form of a bulb, corn, tuber, or fleshy root, which is essentially a food-storing organ. There is no water storage, for these plants are active only during the rainy season. In this sense, they are similar to drought evaders.

As a generalization, there is a gradient of growth forms related to increasing aridity in desert regions. Drought deciduous shrub species tend to predominate in areas of higher moisture availability. As moisture becomes more limited, evergreen shrubs predominate, perhaps because evergreeness reduces the energy costs to the plant of having to replace its photosynthetic apparatus each year. In the very driest habitats succulent species are favored. Since arid environments are often patchy and vary greatly through time and space in relation to moisture conditions, most communities are a mixture of the three growth forms.

Xerophytes have many physiological mechanisms that can contribute to their ability to resist moisture stress. Most can develop low osmotic potentials in their cells (as low as –60 bars) to increase their ability to extract water from the soil. Many xerophytes grow in saline (high salt) soils and have the ability to exude salt from their tissues to allow them to persist in these conditions. Also, various types of biochemical photosynthetic pathways of plants influence their water use efficiency. **C_3** plants tend to be lowest in water use efficiency with **C_4** considerably more efficient followed by **CAM** plants that are most efficient. CAM plants, usually succulents, also have the ability to keep their stomates open at night and closed during the day to allow carbon dioxide capture with minimal water loss, and to use the stored carbon dioxide for photosynthesis during the day with their stomates closed which results in significant reduction of water loss. Plants of all photosynthetic pathways can, of course, physiologically control the opening and closing of the stomates and thus exert control on water loss. All plants must constantly deal with the dilemma of allowing sufficient carbon dioxide to enter the stomates, while limiting the amount of water that is lost. Many adaptations of plants tend to relate to this dilemma.

Hydrophytes

With respect to their relation to water and air, hydrophytes may be arranged into four groups: free floating, submerged anchored, floating-leaved anchored, and emergent. Structural adaptations of hydrophytes are chiefly responses to excessive water content, which, of course, implies decreased oxygen supply. These adaptations consist of a decrease of tissues protecting from water loss and mechanical injury, a reduction in supporting and conducting tissues and a marked increase in aeration of the tissues (aerenchyma) with a corresponding decrease in palisade tissue.

1. Free-floating. Plants in this group (duckweeds, water hyacinth) are in contact with air and water but usually not with the soil. If present roots of these plants are poorly developed and nonfunctional, the entire plant is primarily modified leaf tissue which usually has well-developed aerenchyma tissue, stomata on the upper surface only and a thick waxy covering to prevent clogging of the stomata by water.

2. Submerged Anchored. The roots, stems and leaves of higher plants of this ecological group (pondweeds, water milfoil, bladderwort) are greatly modified. The epidermis of the leaves and stems of these plants are noncuticularized and they are thus able to absorb gases and nutrients directly from the water. The roots are greatly reduced in size, poorly or not at all branched and generally destitute of root hairs. In some species roots have entirely disappeared. The greater density of the water as compared with air renders support less necessary, and the stems are usually long and slender with a poorly developed vascular region. Leaves are finely dissected or thin, linear, or ribbon like and often only two or three cells thick. The thin leaves present an increased surface for reception of diffuse light and provides a large absorbing surface directly in contact with the water. Stomata are absent, except where vestigial, and these are functionless. Air chambers and passages filled with gas are common in all the vegetative organs, i.e., the aerenchyma is abundant. The air spaces not only afford buoyancy but also retain some of the oxygen resulting from photosynthesis to be used in respiration. Likewise, a part of the carbon dioxide that accumulates in these reservoirs at night may be used when the plant is illuminated.

Vegetative reproduction is highly developed and flowers and seeds are less abundant than in most habitats, a response found in most types of aquatic plants. When sexual reproduction in submerged plants does occur, the flowers are usually fertilized at or above the surface of the water.

3. Floating-leaved Anchored. Most of these species (water lilies, pondweeds) develop long, horizontal stems, rootstocks, or tubers. By means of these, they migrate rapidly through the water or mud and, as a consequence, the plants usually form conspicuous communities. Root systems are poorly developed, probably as a result of an excess of water and reduced air supply. The petioles are much elongated. The aerating system is greatly emphasized, and supporting tissues are reduced. Leaves that have developed under water are like those of submerged plants, but in both form and structure, while floating leaves are essentially similar to those of emergent plants. They are usually coated with wax, which prevents the wetting of the upper surface. Stomata are absent with the exception of a few cases where they persist with loss of function on the lower side.

4. Emergent. Plants of this group (cattails, bulrushes, spikerushes, water primrose, smartweeds) are adapted to live partly in water and partly in air. Most species of emergent plants have extensive underground rhizomes or creeping stems, which are rooted in the mud and they spread rapidly by vegetative expansion. Thus, while the roots and portions of the stem, as well as, frequently part of the leaves are under water, at least a portion and often most of the shoot is aerial.

The extent, degree of branching and root hair development of roots varies directly in proportion to decrease in water content. Since both mechanical and conductive tissues are well developed, the plants are able to grow erect without being supported by water. The most distinctive feature of the stems of emergent plants is the large internal air chambers (aerenchyma) crossed by frequent diaphragms that are pervious to air. These are important in affording aeration to the submerged parts.

Leaves of emergent plants show great morphological variation, that is, polymorphism, or phenotypic plasticity when the plant is subjected alternately to the air and water environment. The lower leaves of some species are covered, either normally or by a rise in water level. When they develop under water, they take the form and structure of submerged leaves. The aerial leaves are usually large and entire, showing a marked tendency to increase the exposed surface. As soon as the leaves begin to develop above the water surface, the danger of desiccation is nullified by the production of a cuticle, with its necessary accompanying stomata. Where stomata are present, they have only slightly cutinized walls and in most hydrophytes are almost always open. Stomata are numerous and usually more abundant on the upper than on the lower surface. Aerenchyma tissue is usually well developed.

Temperature

Plants are **ectotherms** in the sense that they rely on external sources for their heat. As a result, plants are often subject to temperature extremes and to survive they must be able to avoid or tolerate those extremes. Plants have evolved various morphological, physiological and life history traits that allow them to cope with their temperature environment. They have some traits that appear to enable them to sense the temperature environment to permit them to forecast future conditions. Many adaptations that appear to deal with the temperature environment are difficult to separate from those that may as well be related to other environmental factors. This is especially true for moisture since temperature and moisture are highly interactive in relation to the physiological functions of plants.

One way plants can avoid the extremes of any environmental factor, including temperature, is to be active only during short periods of the year when conditions are favorable for growth and development. The **annual,** or **ephemeral,** life span is effective in these conditions. Seeds can germinate and plants develop through seed set during short windows of opportunity when temperatures and other environmental factors (e.g., moisture) are favorable and then persist as seeds during periods of stress. As a general rule, seeds are able to tolerate much more severe environmental conditions than the vegetative plant. This approach seems to have developed widely in high temperature, low moisture environments, such as deserts, but has not been selected for in cold, dry areas such as tundras.

Perennial plants, likewise, can avoid extremes by becoming inactive during periods of temperature extremes. Most perennial plants of temperate and arctic regions have this strategy to deal with long periods of cool or freezing temperatures during the winter season. **Quiescence** is cessation of growth imposed and maintained by extreme temperatures. For example, twigs of some woody plants (e.g., *Rosa, Rhus*) continue growth until frost kills the tips and enforces a rest period. **Vegetative dormancy,** on the other hand, is a rest period that is not enforced by low temperature. In most perennial plants, vegetative dormancy occurs at the end of the growing season even though external temperatures remain favorable for growth. Dormancy is then only broken by exposure to temperatures generally below 8°C for extended periods. This may be a mechanism to prepare for cold conditions, which is a form of acclimation referred to as **hardening.** This cold treatment may be a mechanism that plants utilize to sense or measure the cold season so that in nature exposure to cold temperatures followed by warming conditions signals future favorable conditions for growth. The practice of artificially supplying cool conditions to break **seed dormancy** is called **stratification.** Low temperatures are also often required to stimulate flower bud formation. Seeds of many plants of cold regions often require chilling to stimulate germination. Plants also deal with cold temperatures by altering the physio-chemical properties of their cells to avoid frost damage. They may actively transfer water from the cytoplasm to intercellular spaces to keep ice crystallization from harming the protoplasm. They may also produce chemicals, such as glycerol, to lower the freezing point of water and thus avoid freezing.

Vascular plants have evolved various mechanisms for survival in low temperatures, however, they are less well able to persist in the presence of excessively high temperatures. At high temperatures many biological chemicals, e.g., proteins, become denatured which, of course, ultimately destroys the plant. Very high temperatures also, of course, reduce photosynthetic rates and increase respiration rates, which also can become limiting and ultimately lethal to plants. Principal adaptations that protect plants against high temperature injury are: 1) reduction in leaf size and thinness of leaves which allows for more efficient transfer of heat from the leaves to the atmosphere, 2) increased transpiration rates which cools the leaf, 3) vertical orientation of leaves to reduce the direct heat load on the plant—some plants can vary the orientation of their leaves throughout a day or season, 4) white or light colored leaf surfaces or light colored hairs on the leaves to increase the albedo of the leaves, 5) thick, corky bark which insulates the phloem and cambium, 6) and leaf-shedding under extreme conditions. On the positive side, high temperatures are necessary to cause desiccation of the seedcoat of the seeds of some plants to rupture the seed coat to allow germination. The seeds of some desert annuals appear to require high soil temperatures to enhance seed maturation. This preheating treatment is called **sumorization.**

Extremes of temperature are certainly critical problems for plants; however, within normal temperature ranges vegetative growth, flowering, fruiting and germination are most often normal only under alternating temperatures. The response of plants to rythmic diurnal fluctuations in temperature is called **thermoperiodism.** For example, high daytime temperatures favor high photosynthetic rates, while low night temperatures lower respiration rates, which conserve the photosynthate produced during the day.

Light

Plants may be classified ecologically according to their relative requirements of sunlight or shade. Those that grow best in full sunlight are called **heliophytes** (sunflowers, willows, shortleaf pine), and those that grow best at lower light intensities are **sciophytes** (violets, ironwood, red maple, ferns). Among the heliophytes there are some species which, though they grow best in the sun, can grow fairly well under shade. These are called facultative heliophytes, and those sun plants, which cannot do so, are obligative heliophytes. Sciophytes, likewise, can be divided into two groups, depending on their relative ability or inability to tolerate full sunlight.

Classification of plants in relation to shade or sunlight requirement has been of great practical significance in forestry. Trees are categorized according to their relative **shade tolerance.** This enables prediction of approximate successional replacement patterns to be expected within a particular forest region and, consequently, is important in the understanding and management of forests.

Soil

Plants are adapted in many ways to exist in relation to soil factors. Several general groups of plants are recognized. Many species of moist regions grow only on basic soils containing free calcium carbonate or on circumneutral soils from which the calcium has not yet been lost from the colloids. These are called **calciphytes** (e.g., sideoats grama). They disappear from an area as the soils become acid and calcium is lost, or they become confined to springs or stream margins where the soil solution may remain circumneutral. **Oxylophytes** (e.g., blueberries, azaleas) are plants associated with acid soils. They may possibly be excluded from alkaline soils owing to the insoluble state of iron or phosphorus, and at the same time require little calcium. Fabaceae are mainly confined to neutral or alkaline soils because their root nodule bacteria require such environments.

Plants that can tolerate the concentrations of salts found in saline soils are termed **halophytes** (e.g., saltgrass, saltbush, greasewood, glasswort), as compared to the **glykophytes** which occur in nonsaline soils. If not actually dry, these saline habitats may be termed "physiologically dry" because of the high concentrations of salts, which would limit osmotic activity and, consequently, absorption of water by the ordinary plant. The morphological and anatomical characteristics usually appearing in plants of arid regions are common in plants of saline habitats. Succulence is a common adaptation. Yet these xeromorphic characters have been shown to be relatively ineffectual in maintaining low transpiration rates in halophytes. They must then be able to absorb water in spite of the high salt concentrations, and this is possible because of their own high salt contents. Halophytes may be facultative or obligate. Experiments have shown that most halophytes do not require salinity. Because certain species are tolerant to definite ranges of salt concentration and, in addition, to particular salts, they may be useful indicators of soil conditions.

Phreatophyte is the name generally given to plants that send their roots down to the water table or capillary fringe just above the water table, which provides a ready supply of water. Phreatophytes form a definite group of plants but do not belong to any specific family. Their common characteristic is their use of a large supply of water. They have invaded and replaced the native vegetation in many places. Mesquite and saltcedar are important phreatophytes in the southwest.

Fire

Many forests, woodlands, shrublands and grasslands of the world evolved in the presence of periodic fires. In fact, for some ecosystems fire is an essential element that determines their structure and composition. Plants have special features that favor their survival and regeneration under repeated burning.

Some species have unique **growth and development** features. For example, longleaf pine (*Pinus palustris*) has distinct growth stages that confer advantages in regularly burned areas. Following germination, seedlings persist for 5 or more years in a **grass stage** with the apical meristem surrounded by long, fire-resistant needles. After several years, the stem rapidly elongates and raises the apical meristem a meter or more in a couple of years; the fire-resistant needles persist to protect the meristem. This **leader stage** then grows to produce the **mature** tree, which quickly develops a thick **fire-resistant bark,** which is a feature of many fire tolerant pine and oak species.

Many plants are top-killed by fire but are able to regenerate from protected underground structures. Many grasses and shrubs have their meristems located at or just below the soil surface (**hemicryptophytes**) and regenerate quickly following fire; mesquite is a good example. Some species have developed large underground storage tissues with many latent buds (**lignotubers**) from which the plant can regrow.

Plants killed by fire may be favored in terms of regeneration by having hard-coated seeds in the soil seed bank that are stimulated to germinate by the heat and desiccation of the fire which ruptures the seed coat. Woody plants of the genera *Acacia, Arctostaphylos, Ceanothus and Rhus* exhibit this feature and may germinate in large numbers after a fire. Many conifers, particularly pines, produce **serotinous cones,** which do not open to release their seeds until the cones are heated, and dissicated, which in nature is usually accomplished by fire.

It should be noted that many features that predispose plants to effectively deal with fire in their environment also favor them in relation to other factors in their environment (e.g., variable temperatures, drought, herbivory, etc.). Thus, it is difficult to determine whether fire or some other factor may be the selective force that has promoted these features.

Mammalian Herbivory

Herbivores remove parts of, or sometimes all, of a plant in order to obtain their energy and nutrition. Plants have various characteristics that allow them to resist this tissue removal. **Grazing resistance** relates to mechanisms that enable plants to survive in grazed systems. Resistance can be divided into two major subsets, **grazing avoidance,** which deals with mechanisms that reduce the probability of being grazed, and **grazing tolerance** which deals with mechanisms that increase growth following grazing. Avoidance mechanisms include mechanical (spines, awns, leaf/culm ratio, leaf length and position, etc.) and biochemical (secondary compounds that reduce palatability, or directly affect some physiological function of the animal to the extreme of being toxic) characteristics. Tolerance mechanisms facilitate growth following defoliation. Rapid leaf replacement, compensatory photosynthesis, carbon allocation pattern, carbohydrate reserves, etc. are features that influence regrowth after defoliation.

Traditionally, plant species have been classified as **decreasers, increasers** and **invaders** according to their grazing response within a community. How a plant responds to defoliation relates to the combination of avoidance/tolerance mechanisms of the plant, the kind of herbivore and the climatic and edaphic conditions that the plant is growing in.

Decreasers

Decreasers are species in the plant community that decrease in relative abundance under continued moderately heavy to heavy grazing use. They generally are perennials that are palatable to herbivores and, are often the dominant species in the potential plant community (e.g., big bluestem, Indiangrass).

Increasers

Increasers are species that normally increase in relative abundance as the decreasers decline (e.g., Texas wintergrass, buffalograss, three-awn grasses). But increasers do not always react in this simple fashion. Species that increase at first may decrease later as moderately heavy to heavy grazing use continues. Increasers commonly are the shorter, less productive, subdominant species in the potential plant community. Under grazing use, those increaser species of low forage value tend to increase more rapidly than those of high value.

Invaders

Invaders are species that are not normally found within a given habitat because they cannot withstand the competition for moisture, nutrients and light in the climax vegetation. They come in and grow along with the increasers after the climax vegetation has been reduced by grazing. Some are herbaceous annual (e.g., ragweed, horseweed, crabgrass) and perennials forbs (e.g., thistles, ragweeds, camphorweed) and some are woody shrubs (e.g., mesquite, junipers); many have some grazing value but others have little.

Propagule Dispersal

Many plants have evolved specialized structures, which aid in seed transport through the mediums of wind, water or animals. The type of dispersal modification greatly affects the distance and speed of distribution of plants. The agents of seed dispersal and adaptations to these agents are described below.

Wind

1. *Saccate Seed Envelopes.* This modification consists of a saclike membrane, which contains the seed. The membrane provides a light, easily wind-blown container in which the seeds are transported along the ground. Plants of this type are represented by ironwood.
2. *Winged Seed Coats.* These are plants having winged or flattened fruits and seeds and are represented by maple and pine.
3. *Comate Seeds.* Comate fruits and seeds are covered by long silky hair, which increases the surface areas of the seed without appreciable additions of weight. Examples are cattail, willow, cotton, milkweed, and cottonwood.
4. *Parachute Seeds.* Like comate seeds parachute seeds make use of fine, long, pappus hairs. The structural difference is in the suspension of the seed beneath the pappus by a long stem as illustrated by dandelion.
5. *Plumed Seeds.* In this type of seed modification, the style of each seed is modified into a plumelike structure. Examples are clematis and pasque flower.
6. *Tumbleweeds.* The whole aerial portion of the plant is broken loose from its base and swept along the ground by the wind. Tumbleweeds are represented by Russian thistle and witchgrass.

Water

1. *Saccate Seed Envelopes.* Such fruits are impervious to water and contain air chambers in adaptation for dispersal by water. Plants with saccate seeds that may be dispersed by water are sedges and water lilies.
2. *Generally Buoyant Seeds.* Many plants bear seeds, which are capable of floating because the tissues are quite light and buoyant. Buoyant seed parts may include sepals as in some docks, pericarps as in the coconut and testa found in the water flag.

Animals

1. *Awned Seeds.* Most plants having awned seeds are grasses such as needle grass and three-awns. The spear-like tips of awned seeds become imbedded in the fur of animals. Many of these awned seeds are buoyant and are also easily transported by flowing water.
2. *Spiny Seeds.* These seeds are covered by straight spines, which become easily entangled in the fur of animals. An example is sandburgrass.
3. *Hooked Seeds.* The spines of these seeds have tips, which are modified into hooks. Hooked seeds are found on the following plants: cocklebur, burdock, beggar-ticks, and burclover.
4. *Viscid Seeds.* Seeds of this type are covered with a sticky substance and include chickweed, sage, and pinks.
5. *Fleshy Fruits.* Fleshy fruits are transported after ingestion by animals, primarily birds. The seeds themselves pass through the bird's digestive tract undamaged. Examples of these fruits are mistletoe, holly and eastern red cedar.

Man

Man has also become a major agent for plant dispersal around the world. His agricultural, transport, and travel activities offer many intentional and accidental opportunities for seed dispersal.

Mechanical

Some plants (squirting cucumber, woodsorrel, geraniums) have developed morphological adaptations that operate to throw seeds from the parent plant. This is generally short distance dispersal but nonetheless may be effective. Also, by means of vegetative growth of stolons, rhizomes, and other plant parts some species are able to expand their range. Strawberries, Bermudagrass, and cattails are good examples.

Study Questions

1. What are the three major groups of plants in relation to adaptations to moisture?
2. What is the basic distinction between drought evaders and drought resistors?
3. What sort of tissue enhances gas exchange in hydrophytes?
4. Why are the leaves and stems of submerged hydrophytes usually very thin and ribbonlike?
5. Emergent hydrophytes must be able to adapt to two very different environments. What are they and what problems does this cause for these plants?
6. What are obligate hellophytes?
7. What are halophytes?
8. What mechanisms contribute to grazing avoidance in plants?

9. What are the agents of seed dispersal?
10. What kinds of seed dispersal adaptations are most effective for long distance dispersal?
11. What is the influence of fire on seed germination in some plants?
12. What are plants ecototherms?

REFERENCES

Archer, S. A. and F. E. Smeins. 1991. *Ecosystem Processes.* pp. 109–139. *Grazing Management: An Ecological Perspective.* Edited by: R. K. Heitschmidt and J. W. Stuth. Timber Press, Portland, OR. 259 p.

Barbour, M. G., J. H. Burk, W. D. Pitts, F. S. Gilliam and M. W. Schwartz. 1999. *Terrestrial Plant Ecology,* 3rd Edition. Benjamin Cummings, San Francisco, CA. 688 p.

Brown, G. W. 1968. *Desert Biology.* Academic Press, New York, NY.

Cloudsley-Thompson, J. L. 1965. *Desert Line.* Pergamon Press, New York, NY. 86 p.

Daubenmire, R. F. 1974. *Plants and Environment: A Textbook of Plant Autecology.* John Wiley and Sons, New York, NY. 3rd Edition. 422 p.

Larcher, W. L. 1995. *Physiological Plant Ecology.* Third Edition. Springer-Verlag, New York, NY. 506 p.

Mitch, W. J. and J. G. Gosselink. 1993. *Wetlands.* Second Edition. Van Norstrand Reinhold, New York, NY. 722 p.

Handbook of Functional Plant Ecology.

Pugnaire, F. I. and F. Valladares (Eds). 1999. Marcel Dekker, Inc. New York, NY. 920 p.

Pyne, S. J., P. L. Andrews and R. D. Laven. 1996. *Introduction to Wildland Fire.* Second Edition. John Wiley & Sons, New York, NY. 769 p.

Ridley, H. 1930. *The Dispersal of Plants Through the World.* L. Reeve & Co. Ltd., Ashford, Kent. 744 p.

Sculthorpe, C. D. 1967. *The Biology of Aquatic Vascular Plants.* Edward Arnold Pub. Ltd., London. 610 p.

Spurr, S. H. 1964. *Forest Ecology.* The Ronald Press Co., New York, NY. 352 p.

Solbrig, O. T. and G. H. Orians. 1977. "*The Adaptive Characteristics of Desert Plants.*" *American Scientist* 65:412–421.

Weaver, J. E. and F. E. Clements. 1938. *Plant Ecology.* McGraw-Hill Book Company, London. 601 p.

Whelan, R. J. 1995. *The Ecology of Fire.* Cambridge Univ. Press, New York, NY. 346 p.

3.2. Animal Adaptation

Animals exhibit morphological, physiological, and behavioral characteristics that enable them to survive and reproduce in a diversity of environments. These characteristics or **adaptations** represent positive adjustments in structure (morphology) and/or function (physiology and behavior) to a dynamic environment. Individual adaptations of an organism may appear to be relatively "simple" (e.g., a bird feather) but may be only one of a complex constellation of adaptations (structural and functional) associated with bird flight.

As adaptations represent successful adjustments to specific environments, it is philosophically difficult to prioritize these attributes on an importance scale. However, we can identify the procurement of food and locomotion as fundamentally important in the life of all animals. The objective of this exercise will be to examine some of the more apparent external adaptations associated with food procurement and locomotion in animals (specifically in mammals and birds). Examples of characteristics discussed in this exercise will be available via study specimens in the laboratory. Certain of these adaptations can be readily identified in the field by direct observation or through the evaluation of animal signs.

Food Procurement

Adaptations associated with food procurement in mammals and birds may be separated into the problems of (1) food location and (2) food collection.

Mammals

1. Food location. The sensory modalities of vision, smell, and sound represent the most common mechanisms of localizing food sources. Most **nocturnal** mammals (active at night) use vision and smell to locate food while some nocturnal forms use sound. However, with the exception of insect eating bats, smell plays a proportionately greater role in locating food at night. Some whales and dolphins and insectivorous (insect eating) bats use **echolocation** to localize food items. These mammals emit high-pitched, high frequency signals that are reflected off potential food back to the sender. This provides information on the location and movement of the food items. Smell and vision, in particular, are most important for **diurnal** mammals (active during daylight hours) (e.g., antelope).

2. Food collection. Adaptations involved with the direct collection of food items are most apparent externally through examination of the teeth of mammals. The teeth of **herbivorous** (plant eating) mammals (e.g., bison, cattle) are generally characterized by **molars** and **premolars** (cheek teeth) with rectangular-shaped, flat surfaces adapted for grinding plant materials. Herbivores may have sharp, chisel-shaped **incisors** (e.g., rabbits, squirrels) used for cutting coarse vegetation. **Carnivorous** (meat eating) terrestrial mammals (e.g., bobcat, red fox) have prominent, elongate **canine teeth** used to hold and pierce prey. Cheek teeth in carnivores may have sharp, knife-like surfaces that serve to shear meat and bones of prey. Mammals that eat both plants and animals (**omnivores**) have teeth that are structurally and functionally intermediate between those of herbivores and carnivores. The dentition of omnivores includes molars and premolars with rounded surfaces adapted for grinding, canines present but not prominent, and incisors capable of cutting herbaceous plants or the flesh of animals. Such intermediate dentition may be found in man and bears. Certain mammals (e.g., anteaters) do not possess teeth but rather capture ants and other invertebrates with their long, protrusible tongues.

The examination of skulls in the laboratory or from specimens captured in the field can lead to inferences on the general food habits and mechanisms of food getting.

Birds

1. Food location. Vision in birds appears to be the most prominent sense for finding food. This is especially true for most diurnal birds regardless of food preferences. Owls (nocturnal birds of prey) have been shown to use sounds made by moving rodents to localize prey items. Certain birds, such as the American woodcock (Scolopax minor) and common snipe (Gallinago gallinago), locate food in wet soil by the tactile (touch) receptors found in their long, probing bills. The turkey vulture (Cathartes aura) when circling over hidden carrion (decaying flesh) can localize its food through its highly developed sense of smell!

2. Food collection. Birds capture food with their bills, or in the case of birds of prey (hawks and owls) with their feet. Birds do not have teeth (weight reduction advantage related to flight) and hence process food minimally in their mouth cavities and continue the mechanical break-up of food items in their gizzards. The bills (or beaks) of birds vary greatly in size and shape. In fact, the particular bill configuration is closely related to diet.

Graminivorous, or seed eating birds (e.g., cardinals, bobwhite) generally have short, stout and conically-shaped bills. These bills are capable of cracking or slicing large seed hulls. The tongue manipulates and removes the "meat" from the hull prior to swallowing. In some seed eating birds without conical bills [e.g., mourning doves (Zenaida macrovra)] small seeds are swallowed without removing the seed hull. Fish eating, or **piscivorous,** birds (e.g., cormorants, loons, terns) generally have relatively long, powerful, forceps-like bills. In some fish eating birds the bills may have serrations, "tooth"-like projections, or a terminal hook that hold the slippery aquatic prey. **Insectivorous** birds often are characterized by bills with a broad base and a large gape or mouth opening [e.g., purple martins (Progne subis), common nighthawks (Chordeiles minor)]. Such birds take insects on the wing and effectively use their relatively large mouth openings as an insect net. Other insectivores [e.g., black-and-white warbler (Mniotilta varia), red-bellied woodpecker (Melanerpes carolinus), willets (Catoptrophorus semipalmato)] have stout, fine pointed bills. These bills enable the bird to probe into crevices, holes or the soil for insects or other invertebrate prey. Raptors, or birds of prey (e.g., hawks, owls), unlike most birds capture prey (e.g., rodents, rabbits) not with their strong, hooked bills but with their sharp claws or talons. The bill is used after the prey is captured to tear flesh for consumption. **Omnivorous** birds (e.g., bluejays, crows) have generalized bills that can be used to efficiently capture insects, crush fruit, or hold seeds.

Once food has been located and collected it must be processed prior to absorption in the gut. Differences in the anatomy and physiology of the digestive tract of animals are closely related to an organism's food habitats. In carnivorous and omnivorous mammals the relatively simple digestive tract is composed of a pouch-like stomach and small and large intestine. The relatively thin cell walls of animal tissues eaten by carnivores are readily broken down and contents absorbed by this basic digestive tract plan. Herbivorous mammals exhibit a relatively larger and more elaborate stomach and/or intestinal tract. The stomach of ruminants (e.g., cattle, bison) is large and is divided into four separate compartments. This increase in size allows for the fermentation and digestion of cellulose and complex carbohydrates associated with plant material.

Similar differences are found in the gross anatomy of the digestive tracts of herbivorous and carnivorous birds. Birds have a storage compartment (**crop**), a glandular stomach, a muscular stomach (**gizzard**) and an intestine. In most carnivorous birds the glandular stomach is efficient in the chemical breakdown of animal tissues. The muscular gizzard is relatively weak in carnivores but may serve as a trap for bones, hair and feathers. These trapped nondigestible animal parts are often formed into pellets and regurgitated (e.g., owls, hawks). In herbivorous birds (e.g., cardinals, turkeys) the gizzard is powerful and muscular. The gizzard, with the aid of grit (small stones), grinds and breaks up the outer coat of seeds and nuts prior to their movement into the intestine where absorption occurs. In fact, entire walnuts and pecans can be crushed and hulled by the gizzard of wild turkeys (Meleagris gallopavo)!

Locomotion

Movement from one place to another is characteristic of all higher animals at some stage in their life history. Mammal and bird locomotion includes such diverse forms of movement as swimming, running, walking or flight. In some instances the movements of individuals during their lives may be spectacular. For example, the California gray whale (Eschrichtius robustus) travels in excess of 16,000 kilometers (10,000 miles) during its annual journey to its winter feeding grounds in the Arctic Ocean and back again to its breeding grounds off Baja California, Mexico. Similarly, the common barn swallow (Hirundo rustica) traverses over 20,800 kilometers (13,000 miles) in its travels between its wintering ground in Argentina and its breeding sites in North America. In most instances at least portions of the life histories of mammals and birds are carried out in relatively localized environments or **home ranges.** In response to these environments, mammals and birds exhibit characteristic types of locomotion and modes of life. Aspects of these modes of living and associated locomotor adaptations can be examined through observation of living organisms in the field, museum specimens in the laboratory, or by study of animals signs and tracks in the field and in the laboratory.

Terrestrial Locomotion

Most animals and a few birds confine their activities to movement on land surfaces. Mammals are **tetrapods** (have four limbs) and with the exception of certain aquatic forms (e.g., whales) utilize well developed appendages for movement. Some mammals (e.g., man, kangaroos) use only their hind limbs for most movements. The majority, however, employ all four appendages for moving on the soil surface. Certain mammals exhibit **cursorial** locomotion (running). These are typically mammals of grasslands (e.g., bison, antelope, horses). In cursorial mammals the limbs have become lengthened, the number of toes reduced, and the limbs may be moved in one plane only (front to back). Some cursorial mammals [wildebeest (Connochaetes taurinus)] may attain speeds of up to 80 km/h (50 mph) for short stretches. The flightless ostrich (Struthio camelus) of Africa is a cursorial bird capable of speeds up to 48 km/h (30 mph). Certain terrestrial mammals move by a series of short leaps (**saltation**) propelled by the hind appendages (e.g., kangaroo rats, kangaroos). In these organisms the hind appendages are longer and more muscular than the forelimbs. Further, a long, or long and muscular, tail is used for balance. In some predatory mammals (e.g., mountain lion, red fox) that are adapted for chasing prey over short spurts, the appendages may rotate in several planes. This adaptation provides for the great agility necessary for capturing erratic, running prey.

Some birds and mammals are adapted for a mode life that includes some time beneath the soil surface. These organisms are termed **fossorial.** Fossorial birds (e.g., burrowing owls) utilize burrows or tunnels as roosting or nesting cavities. In building burrows, these birds may use their bills or their feet. Mammals such as ground squirrels, kangaroo rats, and prairie dogs live in burrows and forage for plants and seeds found at the surface. In contrast other mammals are completely fossorial (e.g., moles) and are only rarely found on the surface. Fossorial mammals have strong, powerful forelimbs with stout claws ideal for digging. In totally fossorial forms (moles) external ears are lacking and the eyes are **vestigial** (no apparent utility), if not lacking.

Arboreal mammals spend their entire lives in trees. New world monkeys, as examples, have a prehensile tail that assists in balance and movement. Certain arboreal mammals (e.g., tree sloths) have well developed, hooked claws that enable them to cling to branches. Some primates swing from branch to branch with their hands (forelimbs)—a form of movement called **brachiation.** Eastern flying squirrels (Glaucomys volans), found in East Texas, spend a great share of their lives in trees. These nocturnal squirrels are capable of gliding (**volant** locomotion) from branch to branch and tree to tree. Flying squirrels do not fly, but have a membrane connecting front and hind limbs as well as a laterally flattened tail. The membranes and flattened tail enable the squirrel to glide through the air by extending its fore and hind limbs.

Certain birds and mammals spend a considerable amount of time in aquatic environments. Some mammals spend their entire lives in water. These fully **aquatic** forms are adapted for swimming and movement in a liquid medium. Whales for instance, have completely lost the external portions of their hind appendages and move through the water propelled by their tail and flukes. The body of whales is fusiform (spindle) shaped and the front appendages are represented by flippers. Mammals such as river otters (Lutra canadensis), beavers (Castor canadensis), and muskrats (Ondatra zibethicus) on the other hand may be termed **semi-aquatic;** they spend part of their lives on land. These semi-aquatic mammals generally have small ears, fusiform shaped bodies, dense coats of fur, webbed feet (e.g., otter, beavers), and flattened tails. Some birds also spend a considerable amount of time in or on water, usually feeding. Those [e.g., anhingas (Anhinga anhinga), neutropic cormorants (Phalacrocorax brasilianus)] that swim under water after fish use their wings as oars to propel themselves. These forms may also paddle with their broadly webbed feet or use them as rudders. In fact, penguins, no longer capable of flight, use their wings solely for swimming. Waterfowl swim on the surface of the water by paddling with their webbed or partially webbed feet.

Flight, a most common characteristic in birds, is also found among mammals (bats). Both the wings of bats and of birds represent elaborations of the fore-limbs. The wing of the bat is a thin layer of skin spread between the metacarpals and phalanges (what would be bones in the human hand). In birds, the feathered wing is mostly comprised of an elongated radius and ulna (what would be the bones of the human forearm). In bats and birds the bones are light, and a keeled sternum serves as a point of attachment for flight (pectoral) muscles. As most birds do not possess echolocation abilities, the external ears are lost in favor of a stream-lined body. Reduction of body organs (e.g., only one ovary), lack of teeth, extensive air sac and lung system, pneumatized (or hollow) bones, and a body covering of feathers all make birds efficient flying machines.

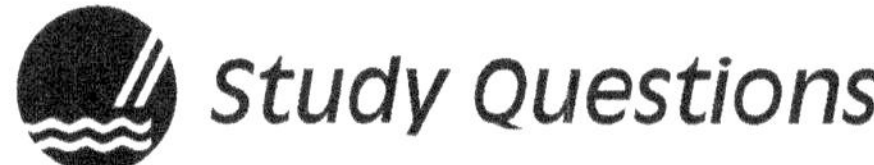

Study Questions

Plant adaptations

1. Define hydrophyte, mesophyte and xerophyte.
2. Distinguish between drought evaders and drought tolerators.
3. List the groups of hydrophytes (1–4).
4. Distinguish between heliophytes and sciophytes. Distinguish between obligate and facultative organisms.
5. Distinguish between calciphytes and oxalophytes. Distinguish between halophytes and glycophytes.
6. What is a phreatophyte? Give an example.
7. Distinguish between decreasers, increasers, and invaders. List methods of seed dispersal used by plants (1–5). Give an example of each.

Animal adaptations

1. Distinguish between morphological, physiological, and behavioral adaptations. Give an example of each.
2. Which of the three senses used by nocturnal mammals plays the greatest role in food location?
3. What is echolocation? Give an example of a mammal that uses it.
4. How do the teeth of herbivores and carnivores differ?
5. Generally, what is the sense most widely used by birds for food location?
6. Besides the beak and feet, what morphological adaptation distinguishes carnivorous and herbaceous birds? How is this feature different in these types of birds?

7. What is a "home range"?
8. Define cursorial locomotion and saltation. Give an example for each.
9. What does the term fossorial mean? Give an example.
10. What is brachiation and volant locomotion? Give an example for each.
11. List adaptations for flight exhibited in birds and/or bats.

REFERENCES

Bolen, E. G. and W. L. Robinson. 1999. *Wildlife Ecology and Management,* 4th Edition. Macmillan, New York. NY. 605 p.

Davis, W. B. and D. J. Schmidly. 1994. *The Mammals of Texas.* Texas Parks and Wildlife Department. Austin, TX. 338 p.

Gill, F. B. 1995. *Ornithology,* 2nd Edition. W. H. Freeman & Co. New York, NY. 763 p.

Martin, R. E., R. H. Pine and A. F. Deblase. 2000. *A Manual of Mammalogy with Keys to Families of the World,* 3rd Edition. McGraw-Hill, New York, NY. 352 p.

Sibley, D. A. 2001. *The Sibley Guide to Bird Life and Behavior.* Knopf, New York, NY. 588 p.

Vaughan, T. A., J. M. Ryan and N. J. Czaplewski. 1999. *Mammalogy,* 4th Edition. Brooks/Cole. 672 p.

Assignment Questions: Plant and Animal Adaptations

Plant Adaptations

Refer to information provided by your TA, and observations made at the Biodiversity Research and Teaching Collection (BRTC). *Online guides can also be a good source of information, but pay attention to the quality of information and, as always, do not plagiarize.*

Leaf traits and aridity

1. Describe a key evolutionary trajectory exemplified by the spines of *Opuntia pheaecantha* Prickly Pear, the leaves of *Larrea tridentata* Creosote bush, and the leaves of *Quercus stellate* Post oak. Are they similar?

2. In which kinds of environments would you expect mosses? Why? What adaptations might be necessary for mosses to occur in variable environments?

3. *Adiantum capillus-veneris* (aka Maiden hair fern) grows all across the state, although the typical habitat for this species is in seeps and very moist areas. How might this plant survive in very arid environments?

4. What are aerenchyma and why might aquatic plants have them?

Seed dispersal

5. Identify the **seed type** (saccate, winged, comate, awned, and hooked seeds, fleshy fruit, etc.) and **seed dispersal medium** (wind, water, animal) associated with the following plant specimens on display:

   pine:

   needlegrass:

   cocklebur:

   beggar lice:

   milkweed:

   cattail:

   rattlepod:

Animal Adaptations

Refer to information provided by your TA, and observations made at the Biodiversity Research and Teaching Collections. *Online guides can also be a good source of information, but pay attention to the quality of information and, as always, do not plagiarize.*

Foraging Strategies

6. Name 2 adaptations that animals use to locate food if they primarily forage during the night?

7. Name one mammal and one bird that have at least one of the adaptations listed in question 2.

8. Why would the ability to locate prey through touch be an advantageous adaptation for the Common Snipe (*Gallinago gallinago*), catfish (Siluriformes) and Raccoon (*Procyon lotor*)?

9. Why is a keen sense of smell advantageous to Turkey Vultures (*Cathartes aura*) for locating food as opposed to good eyesight alone?

10. What morphological traits allow owls to (a) locate prey in the dark and (b) swoop in and catch prey without being detected?

11. How do the teeth of herbivores (e.g., American Bison (*Bison bison*), Pronghorn Antelope (*Antilocapra americana*)) differ from the teeth of carnivores (e.g., Bobcat (*Lynx rufus*), Skunk (*Mephitus spp.*)?

12. Why would it be advantageous to have specialized teeth?

13. What is the advantage of having teeth and beaks that are not specialized, like in omnivores (e.g., Raccoon (*Procyon lotor*), Coyote (*Canis latrans*), Bluejay (*Cyanocitta cristata*), Yellow-billed Cuckoo (*Coccyzus americanus*)?

14. Looking at the beaks of the Northern Cardinal (*Cardinalis cardinalis*) and the American Goldfinch (*Spinus tristis*), what do you think they have adapted to eat most efficiently?

15. Compare and contrast the beaks of Nighthawk (*Chordeiles minor*) and the Yellow-bellied Sapsucker (*Sphyrapicus varius*). Why are they so different considering they are both insectivorous?

16. Compare the bills and legs/feet of Green Herons (*Butorides virescens*) and Brown Pelicans (*Pelecanus occidentalis*), both piscivorous. How are they different and how do these adaptations relate to the way these two species hunt for food?

Locomotion

17. In what environment would it be advantageous to excel in cursorial locomotion?

18. Kangaroo rats (*Dipodomys spp.*) and bullfrogs (*Rana spp.*) both exhibit saltation movement and are a good example of convergent evolution. What morphological traits do they have in common to perform this type of locomotion?

19. Why would it be an advantage to be a fossorial organism (e.g., gopher, mole)?

20. What adaptive trait is apparent in aquatic mammals (e.g., American Beaver, *Castor Canadensis*), amphibians (e.g., Snapping Turtle, *Chelydra serpentine*), and water birds (e.g., Brown Pelican, *Pelecanus occidentalis*)?

21. The adaptation in the previous question is another example of convergent evolution. Why would this be an advantage in an aquatic environment?

4

Population Studies

Population, a group of individuals of the same species inhabiting a given area, is a fundamental unit in ecology and the basic unit of evolution. In addition to being a key component of theoretical ecology, ecological studies of populations have enormous importance in many applied field, from conservation of endangered and threatened species to sustainable development of fisheries, wildlife management, pest management, disease control, and human population growth.

There are several major approaches to population studies: the abundance (number of individuals) of a population and its estimation, the structure (age, size and sex) of the population, and the temporal dynamics (changes over time) of population.

In this section, we are going to examine population sampling techniques for animals (those for plants will be covered in a later section), methods of population estimation, field applications of population estimation, techniques for studying population structure, and system modeling approach for studying population dynamics. We will also examine intraspecific competition, negative interactions among individuals of a population.

REFERENCES

Begon, M. and M. Mortimer. 1986. *Population Ecology: a Unified Study of Animals and Plants,* 2nd Edition. Sinauer Associates, Sunderland, Mass. 220 p.

Ebert, T. A. 1998. *Plant & Animal Populations: Methods in Demography*. Academic Press, New York, NY. 312 p.

Gotelli, N. J. 2001. *A Primer of Ecology,* 3rd Edition. 265 p.

Harper, J. L. 1977. *Population Biology of Plants.* Academic Press, New York, NY. 892 p.

Krebs, C. J. 1999. *Ecological Methods,* Second Edition. Benjamin/Cummings, Menlo Park, CA. 620 p.

Ricklefs, R. E. 2001. *The Economy of Nature.* 5th Edition. W.H. Freeman and Company, New York, NY. 550 p.

4.1. Population Sampling Techniques (Animals)

Sampling of terrestrial animals may involve visual "captures" (sight observations) or the collection of specimens, either dead or alive. The resulting data may provide the raw material for qualitative descriptions of animal populations such as what species are present, extent of home ranges or territories and microhabitat occupied. The collection of specimens may be necessary to maintain an animal in captivity for intensive laboratory investigations. Visual "captures" or the collection of specimens may also provide the information necessary to estimate the numbers of an organism present in a community.

In certain instances it is not possible to capture certain rare or trap-wary species. In fact, sometimes the only clue to the existence of an organism is some **sign** made by or left by the species. If these signs can be related numerically to the number of individuals associated with the sign (1 pair of birds per bird nest), they may serve as an index to population size (see Chapter 4.2). Some of the more commonly found animal signs are: bird and mammal nests, footprints, tail markings, lodges, tunnels, burrows, mounds, gnawed trees, browsed vegetation, feeding debris, runways, fecal droppings or scats and owl pellets.

Although methods available for sampling animals are about as numerous as investigators, certain relatively standard techniques have evolved for sampling the various animal components of a system; these will be briefly described below. The successful utilization of these methods (or equipment) to capture a specific organism must be preceded by a thorough knowledge of the organism's life history. Caution: Before collecting vertebrates, a state collecting permit must be obtained. In the case of most birds, a federal permit is also required.

Sampling Techniques

1. Soil Organisms. The common microorganisms (the most numerous of which are arthropods) of the soil and leaf litter may be collected by separation from their substrate with a Burlese or Tullgren funnel. This is an apparatus in which litter or soil is placed on a fine wire mesh in a funnel below a light source. First, the heat produced by the light and then desiccation drives the soil organisms into a collecting vial. A mixture of 70% alcohol and glycerin can be used in the collecting vial as a killing agent and preservative. Simultaneous collections of soil and leaf litter samples from various microhabitats or different seral stages in biotic succession may thus be compared with changes in arthropod populations.

2. Insects. For diurnal insects, aerial and heavy duty sweep nets may be used to collect insects (and sometimes arboreal amphibians and reptiles) from grass and woody vegetation. Light traps can be used to attract a variety of nocturnal insects. An old sheet with a nearby (overhead) light source will attract the insects that then can be picked off the sheet.

Simple "drop-in" tin can traps may be used to collect ground dwelling insects and arthropods which fall into ("drop-in") the buried cans. The tin can traps may or may not be baited. Collected insects may be killed in a potassium cyanide killing jar or marked (with fingernail polish or enamel) and released for possible recapture.

3. Small Mammals. Small mammals may be collected with poisons or snap traps for museum purposes or for anatomical studies. For mark and recapture studies, live traps are used. Live traps may be purchased (e.g., Sherman Live Traps or Havahart Traps) or they may be constructed. For small rodents various baits can be used, however, scratch grain, peanut butter, and oatmeal work well.

4. Large Mammals. For live trapping larger mammals, snare sets may be used along runs or trails. Various modifications of live traps are often employed (Havahart traps, Tender traps, etc.). The size and the action of traps used depend on the habits and the size of animals to be trapped. The choice of baits (foods or scents) likewise depends on the habits and the food requirements of the target mammals. Some larger game mammals may be successfully herded along "drift" fences that lead into corrals. Leghold traps and poison baits may be used to kill-trap squirrel to dog-sized mammals when dead specimens are necessary.

Track plates may also be used to monitor the movements and presence of predators and other megafauna. These are placed on the ground and are typically comprised of thin acetate sheets affixed to an aluminum (or some other hard backing) flashing covered in a healthy layer of graphite powder. The plates are placed strategically collecting tracks from various "visitors" and can provide a proxy for mammalian diversity or movements in the area. Camera traps are also often utilized and consist of deliberately placed game cameras that continuously record, capture video or still photos at a set time interval, or are triggered by movements of varying degrees. Both track plates and camera traps are relatively noninvasive, benign sampling techniques and so are widely used for a variety of species.

5. Birds. For most purposes, adequate sampling of avian populations may be achieved by visual or auditory "observations." The qualitative description of an avian community can be achieved by noting all of the bird species seen or heard. For quantitative studies, visual or auditory sampling can be related to some standard unit of effort. For example, singing male bird counts consist of short stops at predesignated stations along predetermined census routes or transacts. At these stops counts of singing male birds (presumably territorial males during the breeding season) are made.

Traps that are used to capture birds most commonly are baited or they may involve entanglement of birds in flight. Baited "walk-in" type traps are especially effective for capturing seed-eating birds. Various modifications of these walk-in (swim-in) traps are used to capture game birds such as, mourning doves, bobwhite or waterfowl. Another common baited trap is the cannon-net trap used to capture waterfowl or turkeys baited into a feeding area. When the baited birds enter the feeding area, a large net is projected over the birds with rockets.

Birds that are difficult to catch with baits can often be captured with mist nets. These fine meshes, nylon nets are approximately 12 meters wide by 2 meters high and are placed in flight lines. Birds fly into the nets and become entangled as they struggle to escape. Nets may be placed for passive capture or be "baited" by using a playback device to lure in birds. This technique is most often used for targeting a single species.

6. Amphibians and Reptiles. Amphibians and reptiles may be effectively collected with funnel traps. The funnel traps are set against a long "drift" fence which intercepts and directs the herps toward the funnels. The animals moving along the fence crawl through the funneled openings into the trap and are unable to relocate the opening from inside in order to escape. Large (5-gallon) "drop-in" traps may also be used with a drift fence. These are very effective for the collection of box turtles. In many instances, the successful collection of amphibians and reptiles simply involves the location by sight or sound of individuals. Once found, the individual may be caught with a snare or net at the end of a long pole, stunned with rubber bands, pinned with a snake stick or grabbed with nimble fingers. Examples of the specialized equipment described in this exercise are displayed in the laboratory, or when appropriate, in the field. Notes should be taken on the workings of the traps or techniques, as well as on the organisms for which the methods are most commonly applied.

Study Questions

1. What is an animal sign?
2. For which groups of animals is a "drop-in" trap likely to be used?
3. How does a "drift" fence improve the chances of successfully trapping certain amphibians and reptiles?
4. What are some reasons for live trapping an organism?
5. How does a Burlese funnel work?
6. What must be considered in the selection of bait for a trap?

REFERENCES

Benton, A. H. and W. E. Werner, Jr. 1983. *Manual of Field Biology and Ecology,* 6th Edition. Burgess Publ. Co., Minneapolis, MN. 174 p.

Bibby, C. J., N. D. Burgess, D. A. Hill and S. Mustoe. 2000. *Bird Census Techniques,* 2nd Edition. Academic Press, San Diego, CA. 350 p.

Borror, D. J. and R. E. White. 1998. *A Field Guide to the Insects of America North of Mexico.* Houghton Mifflin Co. Boston, MA. 448 p.

Bookhout, T. A. (Ed.). 1994. *Research and Management Techniques for Wildlife and Habitats,* 5th Edition. The Wildlife Society, Bethesda, MD. 740 p.

Bub, H., F. Hamerstrom and K. Wuertz-Schaefer. 1995. *Bird Trapping and Bird Banding: A Handbook for Trapping Methods All over the World.* Cornell University Press, Ithaca, NY. 448 p.

Burt, W. M. and R. P. Grossenheider. 1998. *A Field Guide to the Mammals,* 3rd Edition. Houghton Mifflin Co. Boston, MA. 367 p.

Conant, R. and J. T. Collins. 1998. *A Field Guide to Reptiles and Amphibians of Eastern and Central North America,* 4th Edition. Houghton Mifflin Co. Boston, MA. 634 p.

Krebs, C. J. 1999. *Ecological Methodology,* 2nd Edition. Benjamin/Cummings. Menlo Park, CA. 620 p.

Peterson, R. T. and V. M. Peterson. 2002. *A Field Guide to the Birds of Eastern and Central North America,* 5th Edition. Houghton Mifflin Co. Boston, MA. 450 p.

Pettingill, O. S., Jr. 1970. *Ornithology in Laboratory and Field.* Burgess Publ. Co., Minneapolis, MN. 524 p.

Sibley, D. A. 2000. *The Sibley Guide to Birds.* Knopf, New York, NY. 544 p.

Smith, R. L. 1980. *Ecology and Field Biology.* Harper and Row Publ., New York, NY. 3rd Edition. 835 p.

4.2. Population Estimation

Basic to the management of wildlife resources is knowledge of the number of individuals present. With certain large, easily seen birds and mammals, population size may be obtained quite easily, as well as accurately, by a **census** (count) of all individuals. For example, each week every whooping crane (Grus americana) that exists in the wild at the Aransas National Wildlife Refuge is counted on its wintering grounds. The whooping crane is a large white bird (approximately 1.3 m in height) that is easily seen in the much shorter coastal marsh vegetation. During the winter months (with snow on the ground and trees without leaves) the moose (Alces alces) population on Isle Royale in Lake Superior can be accurately censused from the air. For most animals, complete enumeration of all individuals is neither practical nor possible. Therefore, it is necessary to make an estimate of the true population size.

One commonly used method to estimate population numbers is to relate some animal sign or object to the population size. Such a sign or object that can be related to a given number of individuals in the population is called a **population index.** Examples of animal signs that can serve as populations indices are: bird nests (1 per pair); beaver pond caches (1 per colony, approximately 7 individuals per colony), deer pellet groups (defecation rate, 12–15 groups per day per individual) and call counts of singing males (1 singing male per pair). The use of indices to arrive at relative population sizes is perhaps the most common method employed by wildlife biologists today. It is obvious from the foregoing that the application and use of population indices must be based on a thorough understanding of the biology of the species and the environment in which it occurs.

A second commonly used technique to estimate population size is to sample the population. For any such estimate, based on samples, to be valid, care should be taken to ensure that the sampling procedures are random and otherwise unbiased.

The size of a population can be estimated with a mark and recapture sampling technique called the **Peterson/Lincoln Population Estimation Procedure (PLPEP).** The initial step in PLPEP is to capture individuals of the animal species of interest. The animals are then marked and returned into the system thus establishing a ratio of marked to the total population of the species. At a later time, usually after a few days, the ratio of marked to unmarked individuals in the population is sampled by means of a second capture. Assuming

$$\frac{M}{T} = \frac{m}{t}$$

where: M = number captured, marked and released in the initial sample,
T = total population size,
m = number of individuals recaptured in subsequent sampling (i.e., marked individuals from first sample), and
t = total number of individuals caught in subsequent sampling (marked + unmarked individuals).

The estimate of the population size may then be determined as:

$$T = \frac{Mt}{m}$$

The initial marking and releasing should be completed at one time (one day) and the recaptures also at one time (another day). However, if a small number of animals is caught during the initial capture, this provision is not feasible. In such a case, captures and releases completed for a period (e.g., a week) may be lumped. The assumption is then made that the initial capture and release was completed at the same time. The interval of time between the first and the subsequent captures should be sufficient to allow for normal dispersion of captured individuals back into the population.

Validity of the PLPEP is based on acceptance of the following assumptions: (1) all individuals in the population have an equal probability of being sampled during either the initial or subsequent sampling events; and (2) there is no change in the ratio of marked (M) to total population size (T) between the initial sampling and the second capture event. Some correlates of these basic assumptions are:

1. No significant recruitment of individuals into the population between capture events should occur.
2. No differential mortality of marked and unmarked animals occurs.
3. The marking procedure should be permanent enough to last from the initial to the subsequent capture event.
4. The marked animals are not trap-happy or trap-shy.

If all of the above assumptions and their correlates can be met, then, and only then, will the population estimate be an unbiased estimate of the true population size. The population size estimate derived from the PLPEP is only an estimate of the true population size. Confidence limits, values above and below the estimate, delimit with a given probability, the range of other acceptable population estimates. For instance, at a given probability of accuracy, say 95% level, an estimate of a rodent population might take the form 30 ± 10. Therefore, any estimate between 20 and 40, for the given confidence level of 95%, would be equally valid. The closer we come to fulfilling the assumptions inherent in the PLPEP the greater the probability that the estimate will be closer to the true population size.

Procedure (PLPEP)

Divide the class into four groups. Each group will work with a hypothetical population, a known number of white beans in a container. The container representing a lake and the white beans representing a population of a fish species (each bean represents an individual fish).

1. Initial sample, mark, and release:
   Capture 25 fish (take 25 white beans out, $M = 25$), mark them (replace them with the same number of dark beans), and then release them back to the lake (container).

2. Second sample and counting recapture:
   Let the fish population mix randomly (shake the container well).
   Capture 25 fish (scoop out 25 beans randomly, t)
   Record the number of fish in the second sample that are marked (dark beans, m) in Table 4.1.
3. Calculate and record in Table 4.1 the population estimates (T) using the formula, $T = Mt/m$.
4. Repeat steps 1–step 3 with:
   - sample size = 50 for both initial sample and second sample;
   - sample size = 100 for both initial sample and second sample; and
   - sample size = 200 for both initial sample and second sample.
5. Use Table 4.2 to pool data from other groups and calculate the class averages and the variance (average of the squared differences between group estimates and class average). Evaluate and rank the accuracy (how close is the average estimate to the true population size) and precision (how variable are the estimates with the same sample size) of the results obtained with different sample size.

   Which sample size would you choose and why (accuracy, precision, and cost)?

TABLE 4.1 Peterson/Lincoln Population Estimation Procedure—Group Results

Observer ______________________ Date ______________________

	Number of Individuals			
Number captured & marked in initial sample (M)	25	50	100	200
Number captured in second sample (t)	25	50	100	200
Number captured in second sample that are marked (m)				
Population estimate (T)				

TABLE 4.2 Peterson/Lincoln Population Estimation Procedure—Class Results

	Population estimates						Rank	
Sample size	Group 1	Group 2	Group 3	Group 4	Class avg.	Variance	Accuracy	Precision
25								
50								
100								
200								

Assignment: Population Studies

1. Peterson/Lincoln Population Estimation Procedure (PLPEP)

   a. What is accuracy?

   b. What is precision?

   c. Follow instructions given by TA and Lab Manual to estimate the "fish" population size and submit your data through Fulcrum. As the data are received, your TA will display them in class

   d. Report the class results here:

Sample Size	Population Estimates						Rank	
	Group 1	Group 2	Group 3	Group 4	Class average	Standard deviation	Accuracy	Precision
25 (a)								
25 (b)								
50								
100								

   e. Which sample size would you choose and why?

4.3. Population and Carrying Capacity

Forces within the biotic community, as well as relationships between the physical environment and organisms, have the potential to alter fundamental ecosystem processes. For example, a pasture that is repeatedly overgrazed by large herbivores will eventually fail to support the population. However, the presence of predators can reduce the number of herbivores, allowing for vegetative recovery and, thus, provide population stability. The carrying capacity of an ecosystem defines the maximum number of organisms that a select area can sustain. Carrying capacity is species-specific and can be influenced by the population density, abiotic factors (e.g. catastrophic storm events, fires, droughts), and anthropogenic disturbance.

The Lesson of Kaibab

The Kaibab Plateau is comprised of approximately 300,000 hectares of aspen, spruce-fir, ponderosa pine, and pinyon-juniper woodland and is located in northern Arizona, bordered to the south by the Grand Canyon. Prior to 1905, the deer in this region were estimated to number around 4000 individuals. Based on environmental surveys, the carrying capacity of the range was then estimated to be 30,000 deer. In late 1906, President Theodore Roosevelt established the Grand Canyon National Game Preserve (later to become Grand Canyon National Park) to protect the "finest deer herd in America".

However, the Kaibab region had already been grossly overgrazed by domestic livestock (i.e. sheep, cattle, and horses) by this time and most of the native tall grasses had been eliminated. To protect the deer, the Forest Service first banned all hunting and then began an intensive extermination campaign to rid the region of known predators. By 1939, 816 mountain lions, 20 wolves, 7388 coyotes, and more than 500 bobcats were killed.

The deer population exploded and by 1920 the range was deteriorating rapidly. The Forest Service responded by reducing grazing livestock permits to increase available forage for the deer and to allow vegetative recovery. However, 3 years later, the vast majority of the deer were still on the verge of starvation and range conditions continued to decline. The Kaibab Deer Investigating Committee recommended that a large portion of the livestock be removed from the region immediately and reopened hunting in an effort to decrease the deer population as quickly as possible. A total of 675 deer were killed the following year, however this was estimated to represent only 10% of the number of deer born that spring. At this point, the deer population was estimated to number 100,000 individuals and over the next two winters, an estimated 60,000 deer starved to death.

Since then, the Arizona Game Commission has strategically managed the Kaibab Plateau region with regulations that balance the needs of livestock ranching and those of the native wildlife populations. Seasonal hunting permits and predator protection efforts are utilized to keep deer numbers in balance with range condition. These tragic winter losses can be prevented by keeping deer population numbers below the estimated carrying capacity of current range.

Optional Assignment: Population and Carrying Capacity

Objectives:

- Graph data on the Kaibab deer population of Arizona from 1905 to 1939
- Determine factors responsible for the changing populations
- Determine the carrying capacity of the Kaibab Plateau

DATA TABLE	
Year	Deer Population
1905	4,000
1910	9,000
1915	25,000
1920	65,000
1924	100,000
1925	60,000
1926	40,000
1927	37,000
1928	35,000
1929	30,000
1930	25,000
1931	20,000
1935	18,000
1939	10,000

1. Graph the deer population data. Place time on the X axis and "number of deer" on the Y axis

2. During 1906 and 1907, what two methods did the Forest Service use to protect the Kaibab deer?

3. Were these methods successful? Use the data from your graph to support your answer.

4. Why do you suppose the population of deer declined in 1925, although the elimination of predators occurred?

5. Why do you think the deer population size in 1900 was 4,000 when it is estimated that the plateau has a carrying capacity of 30,000?

6. Based on these lessons, suggest what YOU would have done in the following years to manage deer herds.

   1915:

   1926:

7. It is a criticism of many population ecologists that the pattern of population increase and subsequent crash of the deer population would have occurred even if the bounty had not been placed on the predators. Do you agree or disagree with this statement? Explain your reasoning.

8. What future management plans would you suggest for the Kaibab deer herd?

5

Aquatic Ecosystem Studies

Water covers approximately 3/4 of the biosphere. It produces a number of habitat types which include streams, lakes, seas and wetlands. The potential variability in the physical, chemical and biological parameters of these ecosystems produces a level of complexity no less than that in terrestrial systems.

Aquatic systems may initially be divided into two basic groups: saltwater and freshwater. Water which contains 35‰ by weight (35 parts per thousand) of various salts is classified as saltwater. The most common salt is sodium chloride, NaCl. The concentration of salts in freshwater, in contrast, is 0.501‰ or less. Brackish waters (estuaries, saltwater bays and saltwater marshes) usually have a salt concentration intermediate between freshwater levels and 35‰. Salt concentrations of brackish waters are highly variable depending on the flux of incoming freshwater, tides and the potential for evaporation.

Freshwater systems may be further subdivided into lentic habitats (standing water) and lotic habitats (running water). Lentic habitats may be called ponds. lakes, bogs, swamps and marshes. Examples of lotic habitats are streams, rivers, creeks, bayous, etc. The science of the study of inland waters (typically freshwater) is called Limnology. Although Limnos literally means lake, present day limnologists study both lentic and lotic systems.

In this section, we will examine the dominant physical and chemical factors and biological factors in the lentic ecosystems; and then contrast them to those of lotic ecosystems.

REFERENCES

Allan, J. D. 1995. *Stream Ecology: Structure and Function of Running Waters.* Kluwer Academic Publishers, New York, NY. 388 p.

Kaill, W. M. and J. K. Frey. 1973. *Environments in Profile: An Aquatic Perspective.* Canfield Press, San Francisco, CA. 206 p.

Mitsch, W. J. and J. G. Gosselink. 2000. *Wetlands,* 3rd Edition. John Wiley & Sons, New York, NY. 920 p.

Wetzel, R. G. 2001. *Limnology: Lake and River Ecosystems,* 3rd Edition. Academic Press, San Diego, CA. 850 p.

5.1. Lentic Ecosystem—Physical and Chemical Factors

In lentic ecosystems, temperature, dissolved oxygen, turbidity, and pH are the most important physical and chemical factors controlling the ecology of these systems.

1. Temperature. The temperature of a substance is a measure of the molecular interactions. The kinetic energy of the molecular interactions is termed heat. The total heat content of an aquatic system is a product of both temperature and volume. Thus, two bodies of water of different sizes and with identical temperatures would have different total heat contents. Nevertheless, the temperature of an aquatic system is more likely to be of direct significance to aquatic organisms than heat content.

As a general rule, temperature fluctuations are not as severe in lentic habitats as compared to those in nearby terrestrial systems. This moderation in temperature fluctuations results from the tendency of water to take up or release large amounts of heat with minimal changes in temperature. Although fluctuations in the temperature regime of a pond or lake are buffered on a daily basis, these systems do exhibit seasonal **temperature stratification.** This stratification or layering may have a significant impact on the distribution of many organisms.

Most of the sun's heat energy (infrared radiation) is absorbed in the upper few centimeters of a body of water. As a result of this differential heating, the surface water molecules expand. The expansion of these molecules results in the lowering of the density of the water of this portion of the water column. The heated waters then float on the underlying cooler waters. As the difference between the upper heated water (**epilimnion**) and the cooler underlying water (**hypolimnion**) increases, the subsequent mixing of the two strata by wind and wave action is prevented. The **thermocline** (or metalimnion) is the zone of rapid temperature change separating the warmer epilimnion from the cooler hypolimnion.

If a system is thermally stratified and a thermocline is present, not only is physical mixing of the layers retarded, but also the thermocline may act as a barrier to various aquatic organisms. In fact, the distribution of various species of fish may be limited by such temperature strata.

Thermal stratification of ponds and lakes in the South may begin as early as March and last until October. In more northerly latitudes, initiation and duration of stratification are correspondingly later and shorter. The onset of cool winds and lower atmospheric temperatures in the fall results in the cooling of the upper surface waters and the subsequent mixing (turnover) of the different temperature layers. As water of the hypolimnion is replaced by cooled surface water, nutrients are brought to the surface—a natural form of recycling of materials. The mixed state lasts from fall until spring when re-stratification begins again as the atmospheric temperatures rise (Figure 5.1).

The configuration of the water molecule is such that its greatest density occurs at 4°C. As the molecules are cooled below 4°C, they become less dense. Therefore, ice floats. In climates in which lentic systems "freeze over" an inverse stratification

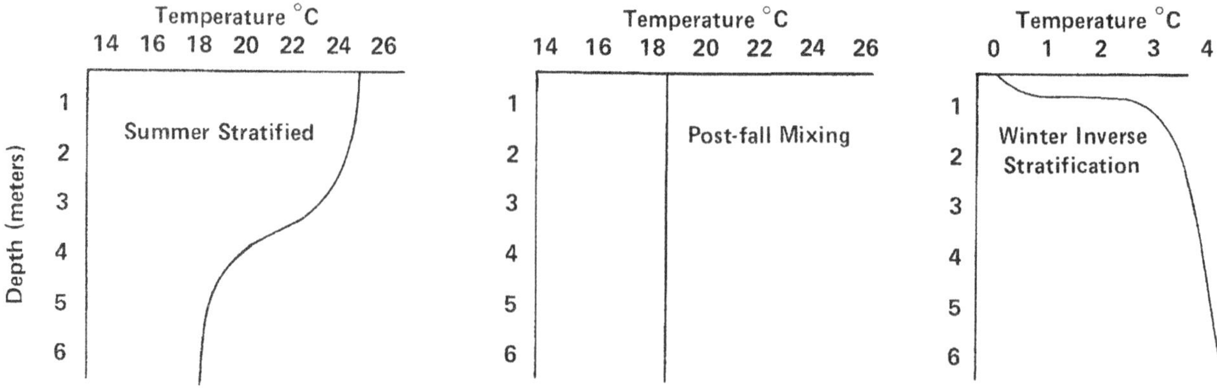

FIGURE 5.1 Temperature profiles of a lentic system at various times of the year.

can occur. That is, the cooler water (from 0°C to 4°C) floats on the warmer waters below. When water freezes into ice, the molecules are spaced even farther apart, thus ice floats even on water at 0°C.

Measure the temperature of a pond or lake at different depths and record your observations in Table 5.1.

2. Oxygen. Dissolved oxygen in a pond or lake is furnished primarily as a result of surface exchange and photosynthesis. Surface exchange occurs via diffusion at the air-water interface and the oxygen is distributed by water currents and agitation. Vascular aquatic plants, attached algae and phytoplankton produce oxygen during photosynthesis. In most cases, the photosynthetic production of oxygen in lentic habitats is the principal source of this gas.

During the period of the year when the pond or lake is thermally stratified, relatively little mixing of the two layers occurs. Consequently, oxygen tends to stratify also. That is, more dissolved oxygen is found in the epilimnion than in the hypolimnion.

The relative amounts of nutrients and organic matter may significantly affect the potential for a lentic system to show oxygen stratification. For instance, if a pond is **oligotrophic** (nutrient poor) oxygen levels may be quite high because the cold hypolimnion can hold relatively more oxygen than the warmer upper layers. However, if a pond is **eutrophic** (nutrient rich), then bacteria and other decomposers in the hypolimnion will utilize the available oxygen. Because mixing is retarded in a thermally stratified system the hypolimnion may become anaerobic. Most small eutrophic ponds and lakes exhibit oxygen depletion in the hypolimnion during the summer months.

As a general rule, dissolved oxygen levels less than 3 ppm are detrimental to most fish. Rapid overturns of stratified waters, which may result from a heavy, cold rain, may seriously deplete oxygen levels throughout the water column by bringing anaerobic water to the surface. Such events may be detrimental to organisms whose distribution is limited to the higher oxygen levels of the epilimnion.

Dissolved oxygen will be measured at different depths with a portable water analysis kit. It is important to avoid air contact with water samples in the dissolved

TABLE 5.1 Lentic Data Sheet—Physical and Cheical Factors

Observer ______________________ Date ______________________

Depth, m	Temperature, °C	Oxygen, mg/l	Dissolved Conductivity, μS	pH

Location: ______________________

Air Temperature: ______________________

Wind speed: ______________________

Cloud cover: ______________________

Time of day: ______________________

Secchi disk Transparency: ______________________

Comments:

oxygen analysis. Instructions for oxygen analysis are found with each kit. Water samples may be obtained from various depths with a **Kemmerer sampler.** Record results in Table 5.1.

3. Turbidity. The depth of light penetration limits the zone of primary productivity in a lentic system. In lakes, light penetrates to the bottom along the lake margin and results in an abundance of rooted aquatic vegetation. This well-lighted shallow water region is called the **littoral zone.** The region of light penetration away from the shore (open water) is termed the **limnetic zone.** The point of effective light penetration—measured by the equality of photosynthesis and respiration—is termed the compensation level. Below the compensation level is the **profundal zone.** The depth of light penetration depends on: light intensity, wavelength, angle of incidence of the sun's radiation, reflectance and transparency.

The factor **transparency** will be examined in this exercise. Transparency is affected by the characteristics of the water itself, by dissolved materials and by particulate matter. The effect of particulate materials is called **turbidity.** The absorption of light by pure water and by dissolved materials varies with the wavelength, but turbidity absorbs light fairly equally at all wavelengths. Highly turbid waters may reduce the depth of light penetration such that the compensation level may be located near the water surface, sometimes within 1 m of the surface. Further, only the upper portions of the water column are heated and the pond stratifies thermally. With reduced light penetration, the depth of the zone capable of supporting plant production is reduced, thus reducing the primary productivity of the entire lake. As many fish rely on vision to identify their prey, high turbidity tends to reduce not only primary productivity but also to reduce secondary and tertiary production.

Ponds are differentiated from lakes based on the extent of the littoral zone. In general, ponds are characterized by effective light penetration to the bottom throughout. Consequently, rooted aquatic plants occur across all areas of a pond. In lakes a portion of the water column lies below the compensation level (profundal zone) and therefore, does not support rooted aquatic plants. In addition, lakes are usually larger than ponds.

Transparency of a lentic system will be measured with a visual method (**secchi disk**) and turbidity mechanically via colorimetry. To operate the secchi disk, lower it into the water column until it disappears from view. Record this depth by noting the water level on the pre-measured line. Lower the disk a few more centimeters and then raise the disk until it is visible again. Record this depth. Average the two readings for the Secchi Disk Transparency (SDT). The SDT (in units of meters, feet, etc.) is useful to compare the transparency of different bodies of water. However, it is necessary to standardize the measurements as the results may be influenced by cloud cover, angle of incidence of the sun's rays and wave action. These variables should be noted when the SDT is recorded. Because the optical characteristics of the disk varies from observer to observer, the same individual should make the SDT determination when sampling a series of lentic systems. Record observations in Table 5.1.

To determine turbidity with the direct reading calorimeter, a water sample is placed in a calorimeter tube positioned between a light source and a photocell. The amount of light attenuated by the sample is compared to a standard. The calorimeter readings are in Jackson Turbidity Units (JTU). This is a measure of the effects of dissolved materials and particulate matter. Water samples may be retrieved from various depths with a Kemmerer sampler. Record your observations in Table 5.1.

4. pH. The pH is measured on a scale of 0–14. A solution with a pH of 7 is neutral, one with a pH greater than 7 is basic, and one with a pH less than 7 is acidic. The pH of a lentic community reflects primarily the geology and soils of the surrounding watershed. Annually the pH of most ponds and lakes remains stable, however, daily fluctuations may be noted. These fluctuations are related principally to the physiological processes of the aquatic plants. In the lighted zones (limnetic and littoral), the lowest pH of the day is reached just prior to sunrise. The low pH is related to the production of CO_2, as only respiration is occurring in the plants during the dark hours. The CO_2 in turn combines with water to form carbonic acid H,CO_3, thus the low pH. As photosynthesis proceeds during daylight hours, the CO_2, produced by aquatic plants and animals during respiration is taken up by the plants and the pH rises again. Determine pH at several depths with a portable water analysis kit. Water samples for different depths from the open water zones may be taken with a Kemmerer Sampler. Record observations in Table 5.1.

Study Questions

1. How can high turbidity affect fish populations?
2. Distinguish between the limnetic and littoral zones of a lake.
3. What are the principle sources of oxygen for a lentic system'?
4. Define compensation level.
5. What factors influence the transparency of water?
6. What is the basic distinction between a pond and a lake?
7. Why does ice float?
8. What causes the daily fluctuations in the pH of a lake or pond.

REFERENCES

Cole, G. A. 1994. *Textbook of Limnology,* 4th Edition. Waveland Press, Prospect Heights, IL. 412 p.

Kaill, W. M. and J. K. Frey. 1973. *Environments in Profile: An Aquatic Perspective.* Canfield Press, San Francisco, CA. 206 p.

Lind, O. T. 1974. *Handbook of Common Methods in Limnology.* C. V. Mosby Co., St. Louis, MO. 154 p.

Smith, R. L. 1977. *Elements of Ecology and Field Biology.* Harper and Row, Publ., New York, NY. 497 p.

Wetzel, R. G. 2001. *Limnology: Lake and River Ecosystems,* 3rd Edition. Academic Press, San Diego, CA. 850 p.

5.2. Lentic Ecosystem—Biological Factors

Physical and chemical factors which play a role in lentic habitats were investigated in the previous exercise. The purpose of this exercise is to investigate the various plants and animals found in this environment. Organisms in a lentic system may be conveniently grouped based on similar microhabitats. The fauna and flora that live in or on the bottom sediments are called the **benthos.** Those relatively immobile and generally microscopic suspended organisms whose movements are controlled by the whims of the current are the **plankton**. Certain macroscopic animals are capable of strong swimming movements and can move voluntarily through the water column. These animals are called the **nekton.** Other plants and animals utilize the leaves and stems of the aquatic vegetation, rocks and other surfaces as a substrate. This group of organisms is called the **periphyton.** Technically this latter term means "attached to plants" but it is used to describe organisms attached to or crawling on any solid surface. Another group of plants and animals is associated with the surface film. These organisms are collectively called the **neuston.** The vegetation of a pond or lake, aside from benthic and planktonic forms, is usually rooted in the bottom and may have floating, submergent or emergent leaves. Specialized equipment and techniques have evolved to sample or measure each of these distinct biotic communities.

Benthos

Benthic organisms may be found in the bottom sediments of the littoral or profundal zones. The benthos are represented by microscopic forms such as bacteria, fungi, single-celled algae and protozoans. The commonly observed clams, snails and crayfish are macroscopic benthic forms. Decomposers (i.e. bacteria, fungi, and some invertebrates) are essential in the mineralization and cycling of various materials. Numerous benthic arthropods are important links in the food chains of a pond—as consumers themselves (grazers, carnivores or filter feeders) and as food for many fishes. Substrate composition (sand, clay, gravel, etc.) is perhaps the most important factor in determining the distribution of the benthos.

Benthos Sampling. The Ekman dredge is commonly utilized to sample benthic organisms. It is portable and easy to use. Because the Ekman dredge is lightweight, it can only be used in systems with soft mud, fine sand or silt bottoms. For coarse gravel bottoms heavier dredges are used. The Ekman dredge is opened and lowered to the bottom by a rope. After contact is made with the bottom the dredge is "triggered" by a messenger and automatically scoops out a sample of the bottom sediment. The dredge should be lowered slowly the last one half meter in order to avoid displacement of the fauna resting on top of the bottom sediments. The contents should then be lifted to near the surface; but not out of the water. Simultaneously place a collecting pan beneath the dredge as the sample is lifted free of the water. This reduces the loss of specimens as water drains from the dredge.

After a sample (or samples) has been obtained with the Ekman dredge, the material is then washed through a series of graded screens or sieves. The mud (the consistency of thick "slurry") is processed through the screen series from the largest mesh to the finest. Each screen retains a particular size range of organisms.

Identify the most common benthic invertebrates in the dredge samples. Use binocular scopes and identification keys provided. Record observations in Table 5.2.

TABLE 5.2 Lentic Data Sheet—Biological Factors

Observer ______________________ Date ______________________

Benthos	Periphyton
Phytoplankton	Zooplankton
Nekton	Neuston
Hydrophytes—free floating	Hydrophytes—rooted floating-leaf
Hydrophytes—emergent	Hydrophytes—submersed

Location: ______________________ Comments:

Air Temperature: ______________________

Time of day: ______________________

Plankton

The plankton consists of both plants (phytoplankton) and animals (zooplankton). These suspended, usually microscopic, planktons have evolved adaptations that maximize their buoyancy. Some of these adaptations are: small size, projections that increase their relative surface area, oil droplets and gas vacuoles. Certain zooplanktons make limited daily vertical movements by swimming or by varying their buoyancy. These diel movements are such that during the daylight hours the planktons may be near the bottom and then rise into the upper levels during the hours after sunset. In large lakes, phytoplankton generally make the greatest contribution to primary productivity. In small lakes and ponds, the rooted aquatic vegetation plays a more significant role. Single species population explosions of phytoplankton are termed blooms.

1. Plankton Sampling. Collection of plankton is commonly made with a plankton net and an attached collecting bottle. The net plankton includes zooplankton and a few of the larger phytoplankton. The commonly used plankton nets (#20 or #25 mesh) collect only 10–20 percent of the phytoplankton. Specific size classes of planktons may be collected by utilizing plankton nets made with different mesh sizes. To sample the small phytoplankton, a centrifuge or a settling technique must be used. Plankton nets may be towed through the water horizontally or they can be drawn vertically through the water column. Planktons often exhibit clumped distributions and/or show irregular seasonal or daily occurrences in lentic systems. Sampling procedures must, therefore, be designed to take into account these irregularities in occurrence and distribution.

Make several plankton tows at different locations in the pond or lake. After the net has been retrieved from each tow, empty the contents of the collecting bottle into a petri dish. Examine the samples with binocular scopes. Identify the most common forms present in the samples. Record these observations in Table 5.2.

Nekton

The nekton is comprised of a wide variety of animals, the most well known are the fishes. Other vertebrates such as various reptiles and amphibians are also represented. The numerous nektonic insects (e.g., water boatmen, giant water bug, mosquito larvae, etc.) play a significant role in lentic environments. Generally, the diversity of the nekton in the littoral zone is greater than in profundal regions. Similar diversity gradients have been noted in the benthos and plankton.

1. Nekton Sampling. Certain of the nekton, such as some adult insects and insect larvae, may be collected with a suitable plankton tow. These insects may also be collected along the shoreline with heavy-duty sweep nets.

Fish and other strong swimming animals are difficult to collect. Techniques for collecting fish vary from the use of seines and nets to electro-fishing and hook and line. The most common method of collecting fish is with a seine. The size (length) of the seine is dictated primarily by the size of the habitat. In narrow inlets and arms of a lake, small seines may be used. Likewise, in larger inlets, or along a stretch of open beach, larger seines or nets must be used. Care must be taken to maintain the lead line (weighted) of the seine on the bottom. If the lead line is inadvertently lifted off the bottom while the seine is being moved through water, specimens may be lost. The most effective seining results from working locations in which fish are naturally trapped or in which they may be "herded" against a barrier.

Seine along the shore and in the inlets of the pond or lake. Record the most common nekton collected in Table 5.2.

Periphyton

The periphyton consist of both plants and animals found attached to, or crawling on, aquatic vegetation, animals and submerged rocks or logs. Diatoms (microscopic yellow-green algae), filamentous algae, protozoans and rotifers are but a few of the commonly found microscopic forms. Snails and some annelids are examples of the larger periphyton. The periphyton are particularly susceptible to environmental degradations. They may be used as indicators of changes in environmental quality.

Periphyton Sampling. The delicate nature of the periphyton makes them difficult to sample and study. These organisms may be sampled by carefully placing a piece of the substrate in a dish and observing it with a binocular scope or by gently scraping them from their natural substrate. A common technique of study is to suspend artificial substrates, i.e., blocks or glass plates in the water. After time for colonization has elapsed, the artificial substrates may be retrieved and the organisms removed by scraping.

Examine with a binocular scope various rooted aquatic plants and/or submerged rocks and debris for the presence of periphyton. Record observations in Table 5.2.

Neuston

The fauna and flora supported by the surface film are called neuston. Protozoans, water mites and springtails (wingless insects) are some of the most abundant animals of the community. The larger water striders (predators) and whirligig beetles (scavengers) are the more commonly seen insect representatives of the neuston. The flora associated with surface film are represented by various algae. Larger forms of the plant neuston are the free-floating aquatic plants including duckweeds and water hyacinth.

Neuston Sampling. The macroscopic neuston may be collected with an aquatic dip net, in Plankton tows or with specialized cone nets. Water samples derived from plankton tows at the surface should yield a host of algae, protozoans and water mites. Record observations in Table 5.2.

Rooted Aquatic Vegetation

In unmanaged ponds and lakes, a distinct horizontal zonation of the rooted aquatic plants of the littoral zone may be observed. In the deep littoral waters, various submerged forms such as bladderwort occur. In shallower waters, floating attached plants (e.g., water lilies) are common. Along the shore the emergent vegetation including cattails and rushes predominate. Eventually, as succession continues, sedimentation will fill in the basin resulting in a bog or marsh.

In ponds and lakes managed for fish, plants in the littoral zone are discouraged because excess vegetation makes the fish difficult to harvest. The diversity of the fauna and flora of the littoral region is consequently low. In ponds with significant populations of rooted aquatic vegetation, the diversity of the biotic community in the littoral zone is high. In certain ponds and lakes free-floating vascular plants [e.g., duckweed (Lemna spp.) and water hyacinth (Eichhornia crassipes)] may cover significant portions, if not all, of the water surface. In such cases, light penetration is reduced and the productivity of the system is low.

Rooted Aquatic Vegetation Sampling. Sampling procedures for the rooted aquatic vegetation may involve both "harvest" and "non-harvest" techniques. In the former technique, samples of the littoral vegetation are harvested (cut or clipped), sorted, dried and weighed. From these data both qualitative descriptions and quantitative information may be obtained. In the non-harvest methods, variations on terrestrial plant sampling techniques (e.g., quadrat, line-intercept) may be employed.

List the rooted plants of the littoral region of the pond or lake studied. Note the horizontal zonation. Record data in Table 5.2.

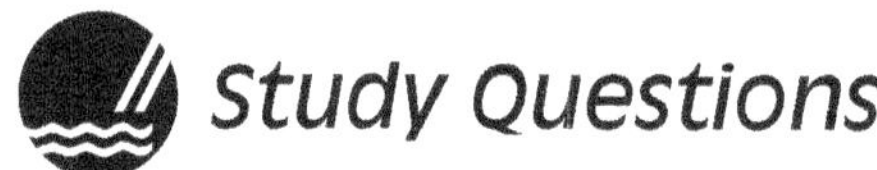

Study Questions

1. Distinguish between the plankton and nekton of a lake.
2. What adaptations of plankton are associated with maintenance of buoyancy?
3. What is an algal bloom?
4. Why is extensive vegetation discouraged in the littoral zone of ponds managed for fish production?
5. Differentiate between periphyton and neuston in a lake.
6. What equipment is used to sample the benthos?
7. What are diel movements of plankton?
8. With a suitable drawing, depict the horizontal zonation of the rooted vegetation of the pond or lake.

REFERENCES

Burton, R. 1977. *Ponds: Their Wildlife and Upkeep*. David and Charles, Inc., North Pomfret, VT. 142 p.

Cole, G. A. 1994. *Textbook of Limnology*, 4th Edition. Waveland Press, Prospect Heights, IL. 412 p.

Kaill, W. M. and J. K. Frey. 1973. *Environments in Profile: An Aquatic Perspective*. Canfield Press, San Francisco, CA. 206 p.

Needham, J. G. and P. R. Needham. 1988. *A Guide to the Study of Freshwater Biology*, 5th Edition. McGraw-Hill, New York, NY. 80 p.

Reid, G. K., H. S. Zim, and G. S. Fichter. 2001. *Pond Life: A Guide to Common Plants and Animals of North American Ponds and Lakes.* St. Martin's Press, New York, NY. 160 p.

Reid, G. K. and R. D. Wood. 1976. *Ecology of Inland Waters and Estuaries.* D. Van Nostrand Co., New York, NY. 485 p.

Welcomme, R. 2001. *Inland Fisheries: Ecology and Management.* Blackwell Science Inc. Boston, MA. 384 p.

Wetzel, R. G. 2001. *Limnology: Lake and River Ecosystems,* 3rd Edition. Academic Press, San Diego, CA. 850 p.

Assignment: Aquatic Ecosystem Studies (I)

1. Plot the temperature and DO profiles; label the axis correctly with measurement unit; draw the profiles for both the lentic and lotic systems on the same charts and label them.

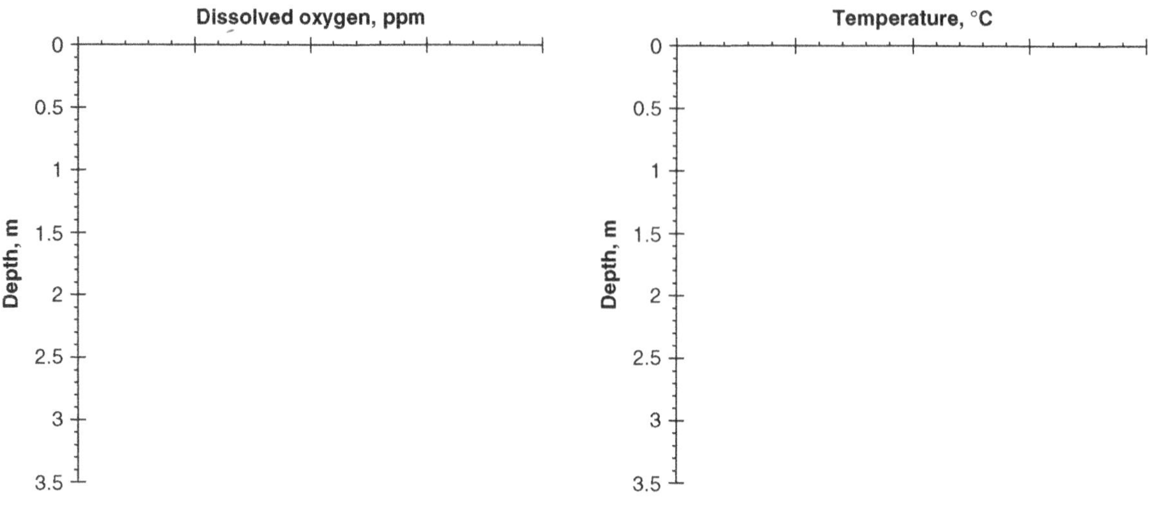

2. For the profiles for lentic system, why did the readings change or not change with depth?

   - Temperature

   - DO

3. <u>How</u> do the temperature and DO profiles for lotic system differ from those for lentic system, and <u>why</u>?

4. Describe any noticeable differences (in kinds and abundance) between the lentic and lotic systems in any of the following categories, and the possible reasons for the differences:

- Benthos and periphyton

- phytoplankton and zooplankton

- nekton

- Hydrophytes (free floating, floating leaf rooted, emergent, submersed)

5.3. Lotic Ecosystem

Streams provide a contrast to the standing waters of a pond or lake. The most obvious difference between lotic and lentic habitats is stream flow or current. Current velocity is dependent primarily on substrate gradient and the width and depth of the streambed. The rate of stream flow varies within the same section of a stream. Water in the center of the channel flows faster than the water on the edges or the slow moving water of a pool versus an adjacent riffle. Further, the water of a stream in hilly or mountainous areas moves much faster than the water of the same stream on the level plains. Headwater portions of a stream, usually with a steeper gradient, are characterized by erosion. That is, erosion of the stream channel exceeds deposition. Whereas, in the lower reaches deposition of materials exceeds erosion.

Because of the current, most of the plankton found in a stream washed in from backwaters, or came from the stream benthos. Rooted aquatic vegetation is not prominent in most streams and rivers. Rooted aquatic plants may occasionally be found in pools and along less turbulent portions of the river margin. The primary productivity of streams and rivers is thus usually less than ponds and lakes because of the reduced populations of producers. The principle source of energy in many streams comes from materials that fall into the stream (e.g., leaves of overhanging trees) or washes into the stream from the surrounding watershed. Therefore, an abundance of detritus and filter feeders are present to provide energy for the secondary and tertiary consumers in the food chains. Here, as in lentic habitats, the composition of the substrate (sand, clay, gravel, etc.) is one of the most important factors governing the distribution of these, principally invertebrate detritus feeders.

The fauna of lotic habitats have evolved various behavioral and morphological adaptations to cope with running water. Many adaptations are associated with reducing the impact of swift currents. These include: (1) avoidance of current by living among the vegetation, under rocks or burrowing into the bottom; (2) streamlined or flattened bodies; or (3) small size, reduction in projecting structures or powers of flight. Other adaptations enable the organisms to maintain their position. Included here are: (1) presence of holdfast or attachment structures such as hooks, suckers or friction pads; (2) production of sticky secretions; and (3) control ballast with sand grains and small stone encasements. Even so, the fast moving currents of lotic habitats carry large numbers of benthic invertebrates downstream. Nymphs of mayflies and stoneflies and larvae of caddisflies and midges are commonly observed drifting in the current. This movement of benthic invertebrates is called stream drift. **Stream drift** is an important source of food for many organisms in lotic habitats and is most abundant at night. Both light levels and temperature influence the occurrence of stream drift. Drift of benthic organisms is most abundant shortly after sunset and least abundant during daylight hours. Further, temperature changes may account for the seasonal variations in drift abundance.

Downstream sections of large rivers typically have lower community diversity than that found upstream. The variety of habitats in headwater sections of a stream (riffles and pools) is generally accompanied by a greater community diversity. The intermittent nature of many small streams has resulted in the evolution of strategies for surviving periods of drought. Encystment, burrowing into bottom mud, aestivation and periodic migrations followed by a rapid recolonization are examples of such strategies.

Streams have a greater ratio of surface area to volume than do ponds and lakes. Because of the relatively greater surface area and because of surface turbulence, oxygen levels in a stream tend to be higher and to show smaller fluctuations. As a result, lotic forms exhibit rather narrow oxygen tolerances. The uniformity in temperature throughout any given section of a stream may be attributed to turbulence, mixing by currents and the relatively shallow waters of lotic habitats.

Physical and Chemical Factors

1. Stream Flow. To determine stream flow, measure and mark any relatively straight, unobstructed (25 to 50 meters) section of a stream. Place a wooden float in the water above the upstream marker. Record the time (in seconds) it takes the float to travel the marked distance. Make at least 3 trials to determine an average time.

Find the average width and depth of the stream. For the width determine the average from at least three points within the marked section. The mean depth can be found from the average of three measurements across the stream at any one point within the marked section. Use the following formula to determine stream flow in cubic meters per second (CMS).

Stream flow (m^3/sec) = WDaL/T

where W = average width of stream.
D = average depth of stream.
L = length of the marked section.
T = average time in seconds for the float to travel the marked distance.
a = constant which varies with type of bottom. Use 0.8 for a rough bottom (gravel, rocks, etc.). Use 0.9 for a smooth bottom (clay, sand, bedrock, etc.).

Record the stream flow in Table 5.3.

The remaining physical and chemical factors will be measured with the same techniques used in the study of lentic habitats.

2. Temperature. Use the thermister telethermometer to record the temperature of the stream at different depths in the center of the stream channel. Record observations in Table 5.3.

3. Oxygen. Measure the dissolved oxygen levels at different depths in the center of the stream channel with the water analysis kits. Record observations in Table 5.3.

4. Turbidity. Use a Secchi Disk to determine the SDT (Secchi Disk Transparency in meters) in the center of the stream channel. The Secchi disk can only be used in sluggish streams. Record observations in Table 5.3.

TABLE 5.3 Lotic Data Sheet—Physical and Chemical Factors

Observer ______________________ Date ______________________

Depth, m	Temperature, °C	Dissolved Oxygen, mg/l	Conductivity, μS	pH

Location: ______________________ Comments: ______________________

Air Temperature: ______________________

Wind speed: ______________________

Cloud cover: ______________________

Time of day: ______________________

Stream flow (CMS): ______________________

Secchi disk Transparency: ______________________

Use the water analysis kit direct reading calorimeter to determine the turbidity from various depths in the center of the stream channel. Record observations in Table 5.3.

5. pH. Measure the pH in the stream with the water analysis kits. Record observations in Table 5.3.

Biological Factors

1. Benthos. Use a screen to sample the benthic fauna from different locations (microhabitats) in the stream. Sort the collected benthos in white enamel pans. Identify the five most common benthic invertebrates from each microhabitat. Record observations in Table 5.4.

TABLE 5.4 Lotic Data Sheet—Biological Factors

Observer ______________________ Date ______________________

Benthos	Periphyton
Phytoplankton	Zooplankton
Nekton	Neuston
Hydrophytes—free floating	Hydrophytes—rooted floating-leaf
Hydrophytes—emergent	Hydrophytes—submersed

Location: ______________________

Air Temperature: ______________________

Time of day: ______________________

Comments:

2. Plankton. Suspend a plankton net in the current of a deep portion of the stream. Empty the contents of the collecting bottle into a petri dish. Examine and identify the most common planktonic forms in the samples. Record observations in Table 5.4.

3. Nekton. Seine as many different microhabitats within the stream section as possible (edges, underbanks, pools, riffles, center of channel, etc.). Record the nekton collected in Table 5.4.

4. Periphyton and Neuston. Make records in Table 5.4 of organisms of the periphyton and neuston observed in the stream. Include as part of your observations the stream microhabitats where these organisms were found (e.g., pools, backwaters, riffles, etc.).

5. Aquatic Vegetation. Note the presence or absence of various forms of aquatic vegetation such as rooted and/or free-floating vascular plants and algae and mosses. List in Table 5.4 the most common forms of aquatic vegetation in or adjacent to the streams.

Study Questions

1. What factor or factors might limit the abundance or occurrence of periphyton in streams? Of neuston?
2. In what way does the production of streams differ from that of ponds or lakes?
3. How does a fisherman utilize a knowledge of stream drift in angling for trout?
4. With respect to stratification (thermal and oxygen) how does the stream environment differ from lentic habitats? Why?
5. Name some strategies that enable organisms of intermittent streams to survive periodic droughts.
6. How does current velocity vary in different parts of a stream channel?

REFERENCES

Allan, J. D. 1995. *Stream Ecology: Structure and Function of Running Waters.* Kluwer Academic Publishers, New York, NY. 388 p.

Cole, G. A. 1994. *Textbook of Limnology*, 4th Edition. Waveland Press, Prospect Heights, IL. 412 p.

Hynes, H. B. N. 2001. *The Ecology of Running Waters.* Blackburn Press, Caldwell, NJ. 555 p.

Kaill, W. M. and J. K. Frey. 1973. *Environments in Profile. An Aquatic Perspective.* Canfield Press, San Francisco, CA. 206 p.

Needham, J. G. and P. R. Needham. 1988. *A Guide to the Study of Freshwater Biology*, 5th Edition. McGraw-Hill, New York, NY. 80 p.

Reid, G. K. and R. D. Wood. 1976. *Ecology of Inland Waters and Estuaries.* D. Van Nostrand Co., New York, NY. 485 p.

Assignment: Aquatic Ecosystem Studies (II)

Step 1: Stream segment length

Measure out a specific length of your stream (if it is a small stream that is moving very slowly, you will probably want to use a shorter length).

Stream segment length: _________ ft

Step 2: Stream segment width

Find the average width of your stream segment at the top, middle, and bottom end of your segment.

Width top: _______

Width middle: ______

Width bottom: ______

Average: ______ ft

Step 3: Stream segment velocity

Using your segment, drop a ping pong ball or a tennis ball (depending on the perceived velocity of your stream-a ping pong ball works better in slower moving water) and record the speed at which the object travels the length of the segment. You should do this at the left, middle, and right side of the stream, and then average your measurements.

Left side (sec)	Middle (sec)	Right side (sec)	Average
Average of all three segments (time in seconds):			

Step 4: Stream depth. Stretch a tape measure across the stream at the mid-point of your stream segment. At 1 foot intervals across the stream, measure the depth (in feet) and record it in the table below.

Interval	Depth
1	
2	
3	

Sum of depths: ______ / number of samples taken = _________ average depth of stream

Step 5: Flow calculation

Now that you have all your measurements, simply plug in the numbers in the equation:

[_____ft (length) x _____ft (width) x _____ft (depth)] ÷ _____time (secs) =_____ cubic feet/sec (cfs)

Community Sampling

Quantitative characterization of plant and animal communities, or populations of an ecosystem, is necessary for understanding and management of ecosystems. This chapter reviews some of the approaches for quantifying community structure. Many of these methods are traditionally used for quantifying plant communities and therefore, the discussion in this chapter focuses on plant communities. These methods, however, can be modified in various ways to sample animal populations, environmental parameters, and other components of an ecosystem.

In the early days of vegetation evaluation, observation and qualitative description were considered adequate to record the vegetational characteristics of an area. This approach is still useful for some purposes. A problem, however, is that few observers see the same things in the same way and few have the ability to translate exactly into words the things they have seen. Thus, the need for more precise vegetation evaluation has led to more quantitative approaches. The keen observer is still a critical factor, but quantitative methods generally provide better records and information that is more readily comparable between observers and from one situation to another. The purpose of this chapter is to provide a baseline of terminology and methodology that will hopefully enable an individual to more fully appreciate the problems of vegetation analysis.

Patterns of Distribution

Populations of plants and animals exhibit distribution in space that may be described as **random, uniform**, and **clumped**. Populations in which the distribution of any one individual is not related to the position of another are randomly distributed. Random distribution of individuals in populations is relatively rare. When individuals are positioned more or less equidistant from one another, the individuals are uniformly distributed. **Territoriality** in animals and allelopathy in plants may result in uniform or even distributions. In many instances, plants and animals are often found clumped or aggregated in portions of available habitat. Clumped distributions in animals are often associated with social aggregations (flocking, schooling, reproduction, etc.) or with aggregations in response to environmental factors (moisture, food, light, etc.). In plants, relatively immobile seeds or vegetative reproduction frequently results in aggregations of individuals.

Measured Variables

Before considering specific sampling methods it is desirable to delimit the kinds of things that are generally measured. Any characteristic of plants can be evaluated, but usually some measure of one of three structural variables is obtained. These variables are: **density**, **dominance**, and **frequency**.

Density is defined as number of individuals per unit area. It answers the question **how many**? It can be determined as follows and is generally expressed per some commonly used unit area such as a hectare (10,000 m^2) or acre.

$$\text{total absolute density of all species} = \frac{\text{total number of individuals of all species}}{\text{total area of sampled units}}$$

$$\text{absolute density of species i} = \frac{\text{total number of individuals of species i}}{\text{total area of sampled units}}$$

These describe **absolute** or **actual density.** For some purposes it is useful to express the contribution of individuals of one species in relation to the total number of individuals of all species. This is referred to as **relative density** and is determined as follows:

$$\text{relative density of species i} = \frac{\text{total number of individuals of species i}}{\text{total number of individuals of all species}} \times 100$$

Dominance is a measure of the size, bulk or weight of the vegetation. It answers the question **how much**? Three characteristics of the vegetation are commonly evaluated as a measure of dominance: **weight** or **biomass, basal cover** or **basal area** and **canopy** (= foliar or crown) **cover** or **canopy area.** Weight measurements, although the best dominance measure, are often difficult and time consuming to obtain and the latter two measures are more often evaluated. The definitions of basal cover and canopy cover are most easily clarified by use of a diagram (Figure 6.1). Basal area is defined as the area of ground surface covered by the bases of plants, while canopy cover is the area of ground covered by the vertical projection of the canopy to the ground surface.

Dominance values can be measured in terms of canopy or basal cover of a species and expressed as a percent of the ground surface covered:

$$\begin{matrix}\text{absolute dominance}\\ \text{of species i}\end{matrix} = \frac{\text{total canopy or basal cover of species i}}{\text{total area of sampled units}} \times 100$$

Dominance can be expressed on a relative basis for each species as follows:

$$\begin{matrix}\text{relative dominance}\\ \text{of species i}\end{matrix} = \frac{\text{absolute dominance of species i}}{\text{absolute dominance of all species}} \times 100$$

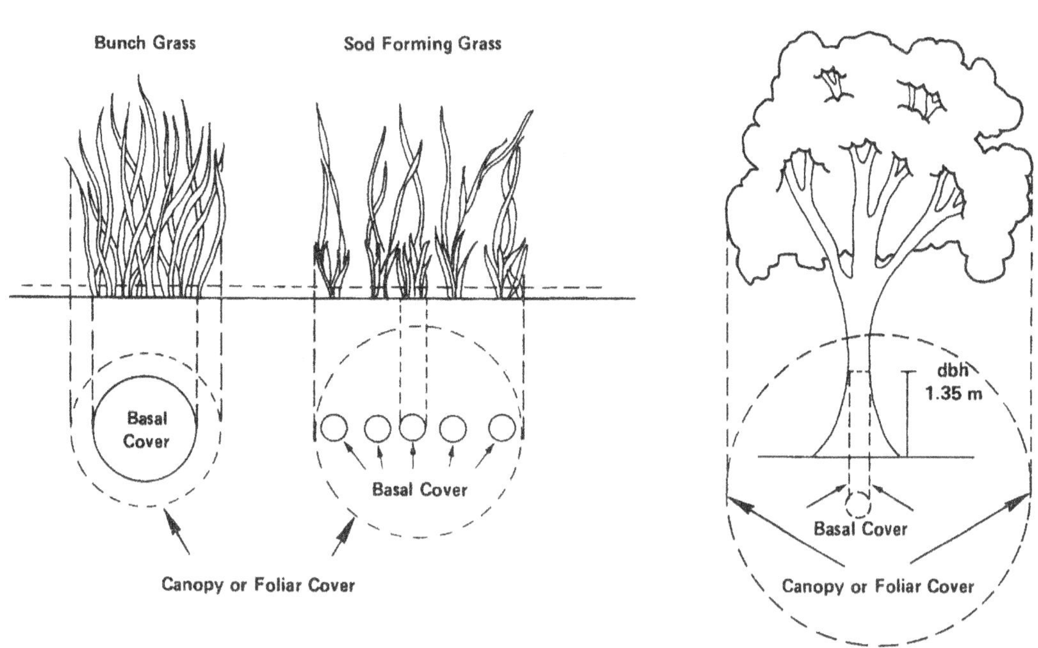

FIGURE 6.1 An illustration of basal and canopy cover.

Frequency is a measure of the commonness and distribution of a species within a study area. It answers the question **where** is the species located and **how common** is it? Frequency is expressed as a percentage of the samples in which a species occurs.

Frequency values are only determined by species and a total for the entire species complement is not calculated.

$$\text{frequency of species i} = \frac{\text{number of sample units in which species i occurs}}{\text{total number of sample units}} \times 100$$

Frequency can also be expressed in relative terms, however, mathematically this is a poorly defined quantity.

$$\text{relative frequency of species i} = \frac{\text{frequency of species i}}{\text{sum of frequencies of all species}} \times 100$$

Take the hypothetical community in Figure 6.2 as an example. Species A, with many individuals spread across the site, has the highest density as well as frequency. Species B, on the other hand, has fewer but much larger individuals and therefore, higher dominance than species A. The relationships for the density, dominance, and frequency of the three species are:

density of species A > density of species B = density of species C,

dominance of species B > dominance of species A > dominance of species C, and

frequency of species A > frequency of species B > frequency of species C.

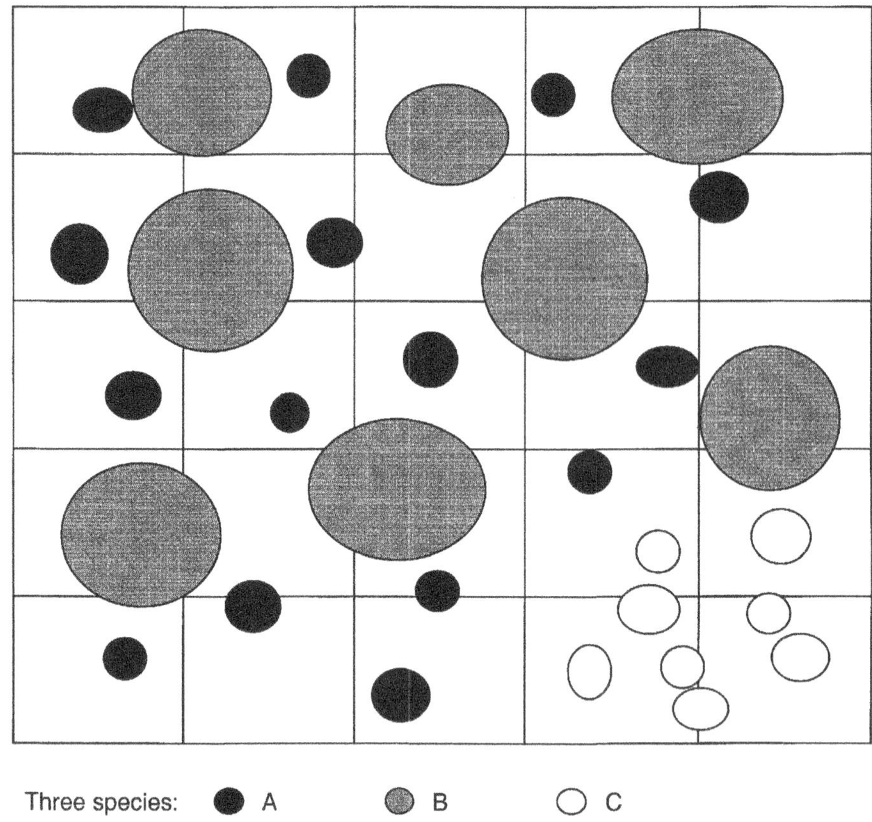

FIGURE 6.2 A hypothetical study area (stand) with three species of plants. Circles represent individuals of species A (black), B (gray), and C (white).

This example illustrates the need to use multiple variables in describing community structures. To simplify the representation, a composite variable called "Importance Value" is sometimes used to quantify the relative importance of species in a community. Importance Value is defined as the average of relative density, relative dominance, and relative frequency:

$$\text{Importance Value of species i} = \frac{\text{relative density of species i + relataive dominance of species i + relative frequency of species i}}{3}$$

Sampling Methods

Density, dominance, and frequency data can be obtained by many kinds of sampling methods. The major categories of methods are:

Plot or Area Type Methods

1. Quadrat
2. Transect

Plotless Type Methods

1. Point Quadrat (an area quadrat reduced to a point)
2. Line Intercept (a transect reduced to a line)
3. Distance (e.g., Point-Centered-Quarter)

Descriptions of these methods are presented in following sections. Criteria and procedures for study area selection, sample location and determination of adequate number of samples are presented in Section 3 that should be consulted in relation to each of the sampling methods. It should be remembered that application of any of the following methods may require flexibility and modification when applied to specific situations.

Study Questions

1. Define the following terms: density, dominance, and frequency.
2. In what ways can dominance be evaluated?
3. A community is sampled and a species is found to occur in seven out of ten samples. What is its frequency?
4. What units are used to express absolute density? Absolute dominance? Relative density?
5. A dominance value is determined for a species to be 5,000 m^2 per hectare. Express this dominance value as a percentage.

REFERENCES

Barbour, M. G., J. H. Burk, W. D. Pitts, F. S. Gilliam and M. W. Schwartz. 1999. *Terrestrial Plant Ecology,* 3rd Edition. Benjamin Cummings, San Francisco, CA. 688 p.

Bonham, C. D. 1989. *Measurements for Terrestrial Vegetation.* John Wiley & Sons. New York, NY.

Cox, G. W. 2002. *Laboratory Manual of General Ecology,* 8th Edition. McGraw-Hill, New York, NY. 320 p.

Curtis, J. T. and G. Cottam. 1962. *Plant Ecology Workbook.* Burgess Publ. Co., Minneapolis, MN. 193 p.

Daubenmire, R. 1968. *Plant Communities: A Textbook of Plant Synecology.* Harper and Row Publ., New York, NY. 300 p.

Elzinga, C. L. and A. G. Evenden. 1997. *Vegetation Monitoring: An Annotated Bibliography.* USDA, US For. Ser. Gen. Tech. Rep. INT-GRT-352. 184 p.

Greg-Smith, P. 1983. *Quantitative Plant Ecology.* University of California Press, Berkeley, CA. 3rd Edition. 359 p.

Kershaw, K. A. and J. H. H. Looney. 1985. *Quantitative and Dynamic Ecology,* 3rd Edition. Arnold, London. 282 p.

Krebs, C. J. 1998. *Ecological Methodology.* Benjamin/Cummings. Menlo Park, CA. 620 p.

Oosting, H. J. 1956. *The Study of Plant Communities.* W. H. Freeman and Co., San Francisco, CA. 2nd Edition. 440 p.

Shimwell, D. D. 1971. *The Description, and Classification of Vegetation.* Univ. Washington Press, Seattle, WA. 322 p.

Smith, R. L. 1990. *Ecology and Field Biology.* Harper and Row Publ., New York, NY. 4th Edition. 922 p.

Weaver, J. E. and F. E. Clements. 1938. *Plant Ecology.* McGraw-Hill Book Co. Inc., New York, NY. 601 p.

6.1. Quadrat Method

One of the most widely used of all sampling methods is the **area** or **plot** sample. This is ordinarily a square area of known size called a **quadrat.** The quadrat method is the most versatile of the plant sampling methods but is also the most time consuming.

Quadrat **size** is generally determined on the basis of the size and density of the plants being sampled. Quadrats should be large enough to contain significant numbers of individuals, but small enough that the individuals present can be separated, counted and measured without confusion. In many forests of the temperate zone, trees are often adequately sampled by quadrats that are 100 m^2 in size, the saplings and shrubs by quadrats 4 m^2 and the herbs and tree seedlings by quadrats 1 m^2. In grasslands, plants are sampled by small quadrats that are usually less than 1 m^2 in size.

While the term quadrat implies a square area, a quadrat may be square, rectangular or circular in shape. Quadrat **shape** is important in relation to the ease of laying out quadrats and to the efficiency of sampling. In low vegetation, circular plots can be laid out very easily by the use of a center pole and a freely rotating radius line. Square shaped quadrats are most often used but, with regard to efficiency, a number of studies have shown that rectangular plots may furnish a more accurate analysis of the composition of a stand of vegetation than an equal number of square plots having the same area. Elongated, rectangular quadrats are called belt transacts.

The **number** of quadrats necessary to adequately sample an area will depend upon the kind of community sampled and the amount of variation encountered. Whenever possible 30 or more quadrats should be used in each stand of vegetation. Any lesser number may give misleading results. Quadrats should be located at random within the sample area.

In each quadrat all plants of the type to be examined are recorded on a previously prepared data sheet. In the case of trees, the basal diameter of each individual tree is measured at a point 1.35 m (4.5 ft) above the ground, which is referred to as **diameter at breast height (dbh).** Canopy diameter or canopy cover can also be measured. For shrubs, saplings and herbaceous plants the number of individuals and canopy cover of each species may be recorded. For the herbaceous layer of a forest, or in a grassland, the grasses and other herbaceous plants may be clipped to determine the biomass of this portion of the community or the foliar or basal cover can be measured. Plants falling on one of the boundary lines of the quadrat should be counted only if more than one-half of the plant is within the quadrat.

The quadrat method will be applied to a forest, woodland or savannah in the field or to mapped populations in the laboratory. Specific details for application of the procedure will be presented in class.

Formulas for analysis of quadrat data:

$$\text{absolute density of species i} = \frac{\text{average number of individuals of species i per quadrat}}{\text{area of a quadrat}}$$

$$\text{relative density of species i} = \frac{\text{total number of individuals of species i}}{\text{total number of individuals of all species}} \times 100$$

$$\text{absolute dominance of species i} = \frac{\text{average dominance (cover or weight per quadrat) of species i}}{\text{area of a quadrat}} \times 100$$

$$\text{relative dominance of species i} = \frac{\text{absolute dominance of species i}}{\text{absolute dominance of all species}} \times 100$$

$$\text{frequency of species i} = \frac{\text{number of quadrats in which species i occurs}}{\text{total number of quadrats}} \times 100$$

$$\text{relative frequency of species i} = \frac{\text{frequency of species i}}{\text{sum of frequencies of all species}} \times 100$$

Study Questions

1. What shape is a quadrat?
2. Would the same size quadrat be used to sample a forest and a grassland? Why?
3. Why would the number of quadrats necessary to obtain an adequate sample vary from one community to another?
4. What is dbh?
5. Are quadrats used to sample only plant populations?
6. How does one obtain the basal area of a species in a community by measuring the diameter of the plants?

6.2. Point Quadrat

Point quadrats are area type quadrats (see Chapter 6.1) reduced to a single point. They are usually made of a 0.5 to 1.0 m long, small diameter pin with a finely sharpened point. As will be observed later, it is necessary to have a sharply pointed pin since blunt tipped pins can produce overestimates of measured parameters. Single pins or groupings of pins (e.g., 10 pins) in a frame can be used for sampling. Research has shown, however, that sampling with single pins is preferred since each pin is located randomly as opposed to a grouping of pins where only the first pin of the group is randomly placed. The pin or pins are usually held in a metal or wooden frame called a **point frame**.

The point quadrat technique is best adapted to communities of moderate to dense herbaceous cover consisting of short plants. It is not useful for sampling of very open vegetation or tall plants such as shrubs and trees. Only dominance (either basal or foliar cover) values can be obtained with this method. Density and frequency cannot be determined.

To apply this method, a pin is pushed down through the vegetation. If basal cover is being evaluated, hits are recorded when the pin intersects a plant base and the soil surface. If foliar cover is being evaluated, all hits are recorded as the pin passes through the vegetation. Some problems exist with this method in that broadleaved plants have a greater tendency to be hit than do linear leaved plants and tall or large plants are more likely to be encountered by the pin than are small, short plants. In addition to recording hits on plants, it is sometimes useful to record hits on litter (mulch) and bare soil to obtain an estimate of the proportion of the sample area characterized by these parameters.

This method generally provides a savings in time and labor over other methods in the analysis of basal area and when used to sample basal area, the method is accurate and free of bias. However, when hits on foliage are used, wind movement and other problems may bias the number of hits and introduce inaccuracy.

The nature of the plant cover, species present and openness of the vegetation will determine the number of samples needed to adequately describe the community. Usually 400–500 points are needed but often many more are required. Unless large numbers of samples are taken only the dominant species of the community are evaluated. Points should be located at random throughout the sample area.

The point quadrat method will be applied to a grassland area in the field or to mapped plant populations in the laboratory. Specific details for application of the procedure will be presented in class.

Formulas for analysis of point quadrat data:

$$\text{total dominance} = \frac{\text{total number of hits on all species}}{\text{total number of pins sampled}} \times 100$$

$$\text{absolute dominance of species i} = \frac{\text{total number of hits on species i}}{\text{total number of pins sampled}} \times 100$$

$$\text{relative dominance of species i} = \frac{\text{total number of hits on species i}}{\text{total number of hits on all species}} \times 100$$

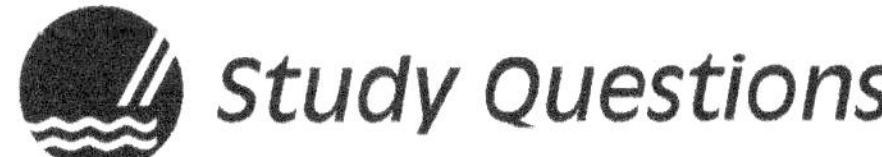

Study Questions

1. What is a point quadrat?
2. What is a point frame?
3. What kind of plant community is best suited to sampling with the point quadrat?
4. What variables of a plant community can be measured with the point quadrat?
5. Why is it necessary to obtain so many samples when using the point quadrat method?

6.3. Line Intercept

A **line intercept** is a transect reduced to a line. A transect, which is essentially an elongated, rectangular quadrat, has a known area, whereas a line intercept is plotless or does not have an area. All standard vegetational measurements may be obtained by this technique, however, absolute density is difficult to evaluate and will not be included here. This technique is useful in sampling non-forest vegetation types such as shrublands, savannahs and bunch grasses, where individual plants do not have overlapping canopies. The method is especially good for measuring dominance but has weaknesses when used to determine density and frequency.

To apply this technique the following procedure is recommended. Several equidistant parallel lines are established across a study area. Along these lines random points are located. A line intercept is laid out at right angles to the compass line and this becomes the sample unit. It is most useful to use a tape measure as the line intercept. With the use of the tape measure scale the line intercept can be divided into intervals of any desired length for determination of frequency. The tape measure scale is also useful for measuring the length of the line intercepted by the canopies or bases of plants. Generally 30 or more intercepts are required for an adequate sample.

Only those plants that are touched by the line intercept or that underlie or overlie the line should be recorded. The length of intercept segments overlying bare ground may also be measured and recorded in the same manner. In sampling communities with more than one plant stratum (e.g., trees, shrubs, herbaceous), it may be necessary to record data separately for each stratum.

The line intercept method will be applied to sample a tree or shrub savannah in the field or to mapped populations in the laboratory. Specific details for applications of the procedure will be presented in class.

Formulas for analysis of line intercept data:

$$\text{relative density of species i} = \frac{\text{total individuals of species i for all intercepts}}{\text{total individuals of all species for all intercepts}} \times 100$$

$$\text{absolute dominance of species i} = \frac{\text{total of intercept lengths of species i}}{\text{total of intercept lengths sampled}} \times 100$$

$$\text{relative dominance of species i} = \frac{\text{total of intercept lengths of species i}}{\text{total of intercept lengths for all species}} \times 100$$

$$\text{frequency of species i} = \frac{\text{number of intervals species i occurs in}}{\text{total number of intervals sampled}} \times 100$$

$$\text{relative frequency of species i} = \frac{\text{frequency of species i}}{\text{frequency of all species}} \times 100$$

Study Questions

1. What is the relationship of a line intercept to a transect?
2. Why wouldn't the line intercept be a useful sampling method in a dense forest in which the canopies of the tree had a great deal of overlap?
3. What unit is used in the line intercept technique to determine frequency?
4. What variable is best evaluated by the line intercept technique?
5. Can basal cover be determined with line intercept?
6. If more than one stratum of a plant community is sampled simultaneously, what influence might this have on the estimate of the total ground surface covered by the plant canopies?

6.4. Distance Measures

A number of sampling techniques have been developed which utilize measurements of distances between plants or from randomly chosen points to the nearest plants for the estimation of density, dominance, and frequency. The most widely used of these is the **point-centered-quarter**, point-quarter, or quarter technique. This technique is best adapted for sampling communities in which the individual plants are widely spaced or in which the dominant species consist of woody plants of shrub size or larger, but it may also work in some open bunch type grasslands.

To apply this technique, a series of random points should be located within the stand to be sampled. The area around each point should be divided into four equal parts, or **quadrants.** This may be done with a compass, or, if the randomly located points are positioned along a compass line, the quadrants may be formed by the line itself and perpendicular to the line at the sampling point. The individual plant nearest the point in each quadrant then is located. The species, basal diameter or canopy diameter, and the point-to-plant distance are determined for the plant and recorded. Point-to-plant distances should be measured to the center of the crown or to the center of the rooted base rather than to the edge of the crown or base. A minimum of 30 points is usually necessary for an adequate sample and even then only the more common species in the community will be sampled. If the community has several layers (e.g., trees, shrubs, herbaceous), it is usually best to sample each layer separately.

This method is very rapid and gives a measure of density, dominance, and frequency. It suffers from two significant limitations. Firstly, in making the calculations for density for the point-quarter technique, the assumption is made that individuals of all species together are randomly dispersed. In most cases where these methods have been tested, this assumption does not appear to produce a significant error. However, in situations where very obvious deviation from overall random dispersion occurs, results obtained should be compared with results from quadrat sampling before extensive point-quarter sampling is undertaken. Relative density and relative dominance values are valid even if the dispersion pattern deviates markedly from random, and will be affected only if they are multiplied by absolute density values in the calculation of the absolute density or dominance of individual species. The second problem is that large numbers of samples are necessary if more than just the dominant species in the community are to be evaluated. Thus, if data are required on all species in the community, the point-centered-quarter may not be as efficient as other sampling methods such as the quadrat.

This method will be applied to a forested or woodland area in the field or to mapped plant populations in the laboratory.

In summarizing the sampling data, point-to-plant distances should first be totaled for all species and all points and averaged to give the mean point-to-plant distance. This value squared gives the mean area per plant. The **mean area per plant** represents the average area of ground surface available to one plant. Total density of plants in the area sampled is then obtained by dividing the mean area per plant into the unit area (e.g., hectare) on the basis of which density is to be expressed. From these calculations and by use of the following formulas, density, dominance, and frequency for each species may be determined.

Formulas for analysis of point-centered-quarter data

$$\text{mean area per plant } (\text{m}^2/\text{plant}) = \left(\frac{\text{sum of all distances}}{\text{number of distances}}\right)^2$$

$$\text{density (plant/ha) for all species} = \frac{1}{\text{mean area per plant}} \times 10{,}000\text{m}^2$$

$$\text{relative density of species i} = \frac{\text{total number of individuals of species i}}{\text{total number of individuals of all species}} \times 100$$

$$\text{absolute density of species i} = \frac{\text{relative density of species i}}{100} \times \text{density of all species}$$

$$\text{absolute dominance of species i} = \text{absolute density of species i} \times \text{average individual dominance (cover or weight) of species i}$$

$$\text{relative dominance of species i} = \frac{\text{absolute dominance of species i}}{\text{absolute dominance of all species}} \times 100$$

$$\text{frequency of species i} = \frac{\text{number of points at which species i occurs}}{\text{total number of points sampled}} \times 100$$

$$\text{relative frequency of species i} = \frac{\text{frequency of species i}}{\text{frequency of all species}} \times 100$$

Study Questions

1. What is mean area per plant?
2. Would the point-centered-quarter method be used to determine stem density in a dense grassland community? Why?
3. What is the major assumption behind the point-centered-quarter method in relation to calculation of absolute density and dominance?
4. What is the sample unit for frequency in the point-centered-quarter technique?
5. What is the biggest advantage of the point-centered-quarter method compared to the quadrat method?

Assignment: Community Sampling

Exercise using artificial sampling boards

Imagine the one m^2 board is really one hectare, and you have just taken an aerial photograph of it from a helicopter. Each dot on the board is a plant and you have coded different species with different colors. Now let's try to characterize the community on this one hectare of land with different sampling methods.

Conduct the sampling in 4 groups; each group uses both random and cluster boards; each group does all methods. Use the appropriate table in the Fulcrum app for each method. Get the data from all groups before you do the calculations.

When you finish sampling and calculation, discuss the following:

1. Compare results obtained with different methods and compare them to the actual data (artificial population board statistics). Discuss when each method works well.

   absolute density   percent cover (dominance)

2. What are the reasons for deviations from actual data?

3. Is the deviation less on random board than cluster board? Why?

4. Little bluestem (*Schizachyrium scoparium*) is a bunchgrass that typically occurs in the savannah. Which method would be best for determining the relative dominance of little bluestem compared to other grass species in a given area? Why?

5. Post oak (*Quercus stellata*) is a tree common in the savannah. Which survey method would be most appropriate for determining the frequency of post oaks? Why?

6. What survey method would be most appropriate for determining the density of herbaceous species (e.g., Cherokee sedge (*Carex* cherokeensis)) in the woodland? Why?

7. Yaupon holly (*Ilex vomitoria*) is a shrub that occurs in clumps, typically as an understory species. Which survey method would you use to determine density of yaupon holly? Why?

Artificial Population Board Statistics:

	Random Population		Cluster Population	
Parameter	All dots	Red dots	All dots	Red dots
Absolute density, #/m^2	1047	157	1840	153
Percent cover, %	12.0	1.66	13.0	1.1

7

Ecological Survey and Study Design

Purpose

One of the major problems in ecology is the determination of population distributions, sizes, and how these change over space and time. Many variables interact to determine the presence or absence of plant and animal species and their abundance. The local distribution and abundance of plant populations is controlled largely by climatic factors, soil factors, drainage regime, land use, and animal influences. Animal populations are likewise controlled by climate, soil, moisture regime and land use, but often, more importantly, by the structure and composition of the biotic community.

It is often necessary to quantitatively describe variations in populations in order to document changes that may occur through time and space, or as a result of a treatment applied to an area in the context of an experiment. The purpose of this section is to discuss terminology, procedures and some common problems associated with the quantitative survey and analysis of plant and animal populations in the field.

In order to describe the relationships between plant species abundance and distributions, ecologists utilize a variety of vegetation survey methodologies. Once characterized, the structure of these plant communities can be compared with variation in soil or other factors that are hypothesized to control it. From the standpoint of land management, range scientists, foresters, wildlife biologists, agronomists, soil scientists and others need to know the kinds and quantities of plants in a community to determine: (1) the amounts of forage available for wildlife and livestock, (2) the amount of timber present, (3) the wildlife habitat present, and (4) other characteristics of the community.

In this chapter, we will introduce several popular vegetative survey methods and discuss experimental design as it relates to sampling plant and animal communities.

7.1. Experimental Design

Every ecological study is built on a well-defined objective. Outcomes of the study are then related back to the initial objectives to determine if the results support the researcher's initial expectations, if any inferences may be made about larger or similar systems, or if the observations lead to surprises that might motivate additional research. This sequence is referred to generally as the **hypothesis-driven** approach and significant progress in ecology today is the result of this approach. Ecologists must address several issues prior to designing a study (each will be addressed further in this chapter):

1. Establish a study subject – what are we studying? (e.g. single animal species, vegetation community)
2. Define objectives/hypothesis (what question(s) are we trying to answer?)
3. Define the geographic area (to what spatial extent do our objectives apply?)
4. Delineate where samples will be gathered and bound the study length in time
5. Outline relevant metrics or parameters necessary to address objectives
6. Determine the ideal survey or sampling techniques within the constraints of the study question, time, and resources available

Study Subject and Objectives

It is important early in any ecological study to clearly define the biological entity you want to make an inference about. We can, for example, make the distinction between studying **wildlife** and **wildlife habitat**. Studies aimed at answering questions dealing with wildlife are often narrowly focused and sometimes physiological in nature (e.g. metabolic regulation in Ruby-throated hummingbird, gender ratios in a White-tailed deer population). However, if your study is directed at making an inference about the ecological processes that structure wildlife communities, your subject may be wildlife habitat. For example, in a study assessing the impacts of grazing from introduced Axis deer in Texas, sampling efforts would be directed at vegetative characteristics (the Axis deer habitat) and not necessarily the deer themselves (especially, for example, if you already had deer population estimates). It is important to remember also that "habitat" is species specific. That is, although many species might overlap in some of the key characteristics of their habitat, they will also differ from each other in some factors. Habitat is therefore not a synonym for vegetation type. Thus within a vegetation type, there can be many species of animals, each of which occupies at least a somewhat different habitat. This distinction is critical when we then want to manage an area for multiple species; think managing for elk and salamanders that both live within a forested area.

Clear, concise objectives are crucial to sound study design. Even monitoring and inventory studies require precise objectives defined *a priori*. Ecologists have many resources from which to gather information necessary to define relevant study questions. The scientific peer-reviewed literature is the most relevant of these, since it represents what is understood by the scientific community at large. Literature searches not only help researchers understand the nature of their target species or system, but also guide them in developing relevant research questions by highlighting missing information or peculiar observations. Research objectives should be

SMART (specific, measurable, achievable, realistic, and time bound) and typically begin with terms like *determine, examine, investigate, improve*, or *evaluate*. Objectives may vary with type of study but ultimately are based on the formal scientific context from which the study is derived and to which the study will ultimately contribute.

Identifying Study Area and Variables

There is no universal method for defining the location of your study. However, the site (or sites) should be relevant to the study subject, time constraints, and accessibility. Placement of locations used for gathering data (i.e., study sites) should also reflect the conditions across the geographic area for which you want your results to have relevance (applicability). For example, gathering your data only near easily accessible roads along a river will be unlikely to reflect conditions on the adjacent—and steep—hillsides.

An intensive survey, or **reconnaissance**, of the potential study area(s) is often performed to become familiar with the biota and physical characteristics of the site(s) prior to conducting a study. The size of the study site(s) should also be chosen with consideration given to the initial research objectives – the size of the study site(s) should be suitably large to encompass enough environmental variability to adequately represent the study subject or, likewise, suitably small to allow sampling any micro- flora or fauna to be included in the study.

Thus, the size of the study site(s) and the environmental variables to be measured are inherently connected. Although often used interchangeably, we should make the distinction between environmental parameter and variable. An environmental **parameter** is any measurable property whose value is a determinant of the characteristics of an ecosystem. An example may be the number of Post oak trees inside Lick Creek Park or the moisture content of the soils in Lick Creek Park above a certain elevation. Chosen parameters should be relevant to the study objectives. For example, if your objective is to estimate the relative abundance of Yaupon holly in Lick Creek Park, counting Yaupon holly separately and counting all other species as one group may be a good idea. Even inside one park, measuring and acquiring values for the parameters directly (e.g., individually counting all trees) is often prohibitively time-consuming. So, we can select locations within the park, measure the soil moisture content or count the number of Post oak trees at these locations and then statistically infer the total from those observations (e.g., # of trees per unit area of the locations multiplied by the total area of the park). That value is a **variable**, or the statistical representation of the true environmental parameter.

Sampling Techniques

Examples of some sampling techniques were outlined in Chapter 4. However, it is important to note that not all techniques are necessarily appropriate for their target species or communities in all studies. Consideration should be given to the time or accessibility constraints (e.g. public/private land, requirement of an Animal Use Permit) as well as any pre-existing knowledge of the animals or plants in the area. For example, Lick Creek Park encompasses a variety of vegetative types. If you were to use 'drop in' or pitfall traps (traps that rely on animals falling into holes in the ground) to sample for small mammal presence, this technique would be appropriate for the upland sites but could be ill-advised in certain bottomland sites due to the risk of flooding. There, Sherman traps (baited box traps with a trigger

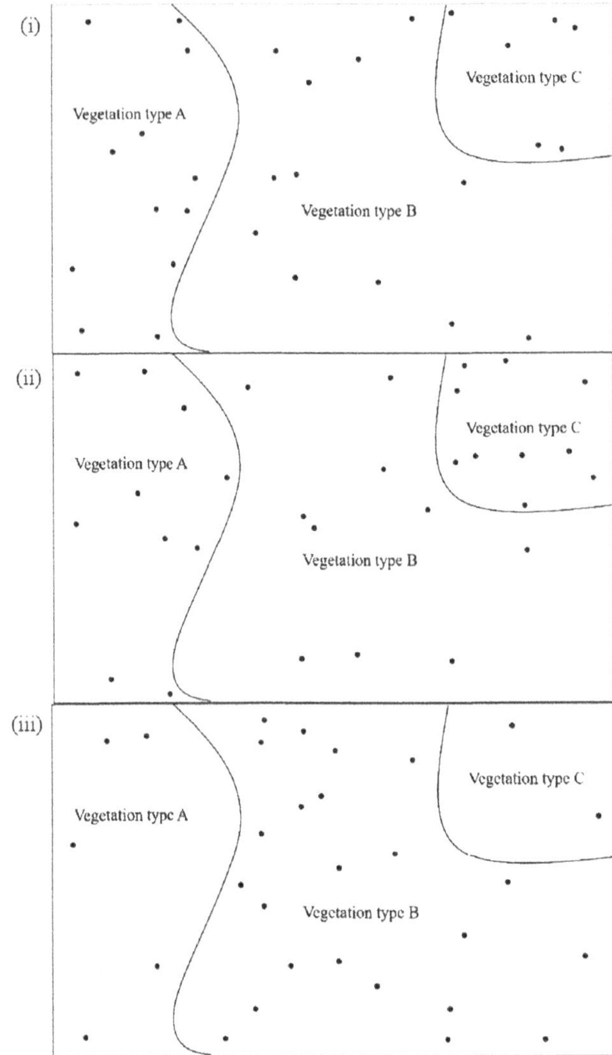

FIGURE 7.1 An illustrated example of simple random (i), proportional stratified random (ii), and disproportional stratified random sampling (iii). The simple random plot shows 30 sampling points distributed across the study area at random, with no consideration given to vegetation type. The proportional stratified random sampling plot shows 30 sampling points distributed at random within each vegetation type proportionally (10 points per vegetation type, regardless of size). The disproportional stratified random sampling plot shows 30 sampling points distributed at random within each vegetation type according to the size or expanse of the vegetation zone ensuring that no group is under- or overrepresented.

that closes the animal inside) may be more appropriate. Additionally, you should consider the distribution patterns discussed in Chapter 6 of your study subject(s). **Random sampling** is common in ecological studies as it eliminates sampling bias and is rather straightforward mathematically, but it can pose challenges:

1. True randomized locations tend to be more clumped and patchy than expected.
2. In studies with small sample sizes, entire strata or groups may be under- or overrepresented.
3. Organisms are rarely distributed randomly.

Stratified random sampling may be ideal in some studies, depending on the study sites and distributional patterns of your study subject(s). Stratified random sampling involves placing sampling points or plots proportionately or disproportionately per defined **strata** (e.g. vegetation type, elevation zone) (Fig. 7.1).

REFERENCES

Morrison, M. L., Block, W. M., Strickland, M. D., Collier, B. A. and M. J. Peterson. 2008. *Wildlife Study Design*. Springer Science & Business Media.

Friedland, A. and C. Folt. (2009). *Writing Successful Science Proposals* (2nd ed.). New Haven, CT: Yale University.

Hailman, J. P. and K. B. Strier. 2006. *Planning, Proposing, and Presenting Science Effectively: A Guide for Graduate Students and Researchers in the Behavioral Sciences and Biology*. Cambridge University Press.

Ricklefs, R. E. and G. L. Miller. 2000. *Ecology*. New York: W. H. Freeman and Co., 23–24 p.

7.2. Lick Creek Park (LCP) Reconnaissance

Here, we will do an intensive survey of potential study sites at Lick Creek Park in College Station, Texas. Careful consideration will be given to the gradients in vegetation and soil characteristics (soil moisture, clay, and sand content, plant species and growth form) with respect to soil water availability.

Background

Lick Creek Park (LCP) is comprised of 515 acres (approximately 208 hectares) of Post Oak Woodland located in the Lick and Alum Creek floodplain, a major tributary to the Navasota River. Historically, it was privately owned and heavily grazed before it was acquired by the City of College Station in 1987. Since then, it has been managed as a city nature park and used for public and educational purposes. The park contains well-developed riverine and alluvial hardwood forest, oxbow meadow, upland post oak savanna, and sandy prairie vegetation types, and maintains a relatively high level of species diversity. We will be focusing our survey efforts on three vegetation types: the upland forest and savanna and the bottomland forest.

Upland Forest

The upland forest dominates much of the area of Lick Creek Park away from the lowlands and creeks. Much of this forest was historically likely more open, savanna and grassland-type vegetation, but has come to be dominated by woody species in part because of fire suppression and historical grazing. Dominant woody plants include trees such as post oak (*Quercus stellata*) and winged elm (*Ulmus alata*) with other oak species like black-jack (*Q. marilandica*), water (*Q. nigra*), and willow (*Q. philos*) being less common. In the understory, yaupon (*Ilex vomitoria*) is common, as is American beautyberry (*Callicarpa americana*).

Upland Savanna

The upland savanna area in LCP is an herbaceous matrix with woody clusters. It represents a historic mid-grass prairie dominated by little bluestem (*Schizachyrium* scoparium, herbaceous layer), yaupon holly and farkelberry (shrub/sapling layer), and scattered individuals or cluster genreally dominated by Post oak (canopy layer). Soils are typically shallow, sandy loam over a clay pan parent material and drainage is generally good. Complex interactions between grazing and fire have affected these vegetation types historically, but in general with increased grazing intensity and lower fire frequency (as a result of fire suppression) the proportion of woody vegetation increases relative to the herbaceous component. Table 7.1 lists plant species for Lick Creek Park and identifies their commonness by vegetation type.

Bottomland Forest

The bottomland forest area in LCP is marked by a significant change in species composition from the upland forests and savanna through mesic forests transistions to relatively wet bottomlands. This vegetation type is densely vegetated with a higher water availability and frequent seasonal flooding. Sycamore, Sugarberry, Water Oak, and Elms dominate

TABLE 7.1 Checklist of plant species found at LCP (A=abundant, C=common, R=rare)

Common name	Scientific name	Upland Savanna	Bottomland Forest
Overstory			
Post Oak	*Quercus stellata*	A	R
Black-jack Oak	*Quercus marilandica*	A	R
Winged Elm	*Ulmus alata*	C	R
Water Oak	*Quercus nigra*	C	A
Willow Oak	*Quercus phellos*	R	A
Cedar Elm	*Ulmus crassifolia*	R	A
Sycamore	*Platanus occidentalis*	R	A
Sugarberry	*Celtis laevigate*	R	A
Honey Locust	*Gleditsia triacanthos*	R	C
River Birch	*Betula nigra*	R	C
Understory			
Yaupon	*Ilex vomitoria*	A	C
Farkleberry	*Vaccinium arboretum*	C	C
American Beautyberry	*Callicarpa americana*	A	R
Eastern Red Cedar	*Juniperus virginiana*	C	R
Poison Ivy	*Rhus radicans*	C	C
Willow Baccharis	*Baccharis spp.*	R	R
Spanish Moss	*Tillandsia usneoides*	R	C
Herbaceous			
Chasmanthium	*Chasmanthium spp.*	R	C
Little Bluestem	*Schizachyrium scoparium*	A	R
Oldfield Threeawn	*Aristida oligantha*	A	R
Brownseed Paspalum	*Paspalum plicatiulum*	A	R
Texas Croton	*Croton texensis*	C	R
Sedge	*Carex spp.*	R	A

the canopy layer with abundant epiphytes. Although there is less light availability at ground level, the herbaceous layer still exists in a sparse to moderately dense floor of Chasmanthium and sedges. Soils (sandy loam to loam) are deeper here with no clay pan. Table 7.1 lists common (and rare) plant species for this vegetation type.

Soil Characterization

Soils have many physical and chemical properties that can be used to describe, classify and relate them to organism distribution and abundance. Only a few of the readily observed or measured soil properties are considered in this exercise. Most characteristics presented here can be used as general surrogates to represent more direct environmental factors such as temperature regime, moisture availability.

For comparative purposes between sites, the overall depth of the soil (solum) is an indirect measure of soil volume, and hence soil moisture storage capacity for plants. Also, overall depth relates to burrowing depths for ground dwelling animals. Depth and color of each horizon (A, B, C) provides information on soil formation processes, environmental relationships and land use practices (Fig. 7.2). Depth can easily be determined by using an incremented, small diameter metal rod and pushing it into the soil at several locations across the site until it reaches resistance due to presence of parent material, a clay pan, rocks or other obstruction. A soil auger could be used to accomplish the same objective. Better yet, a small pit can be dug with a spade to determine the depth of soil, as well as, to observe and measure the depth and color of the individual soil horizons along with other features of the soil such as rockiness, structure, etc. Color of the soil horizons can be generalized and described by the primary colors, or a Munsell color chart can be used to obtain a quantitative description of the soil color.

Soil texture is an important feature relative to plants because it is a major factor in determining the moisture and nutrient relations of the soil as a growth medium and it influences rooting patterns. Texture is a determinant of the kinds and locations of various soil inhabiting microorganisms and burrowing organisms within the soil. Texture can be approximated by the feel method as described in Figure 7.3. More quantitative techniques are available to provide specific percentages of sand, silt and clay for a more exact assessment of texture if necessary. These techniques are described in most soil science textbooks. For many ecological survey studies the feel method, with some practice, is sufficient.

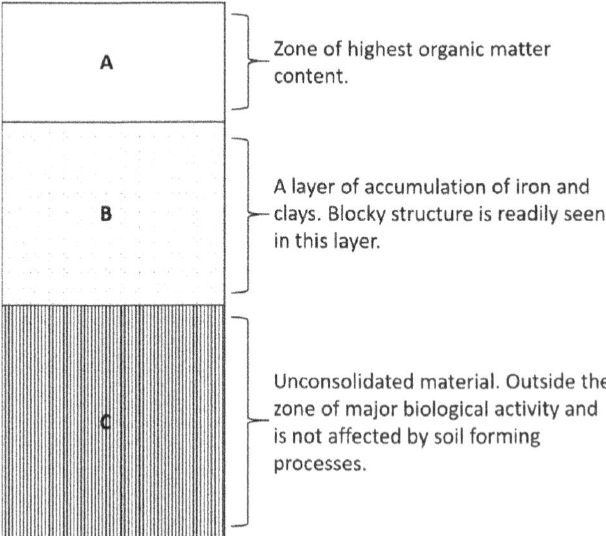

FIGURE 7.2 Soil vertical layers (A,B,C) illustrated with defining characteristics.

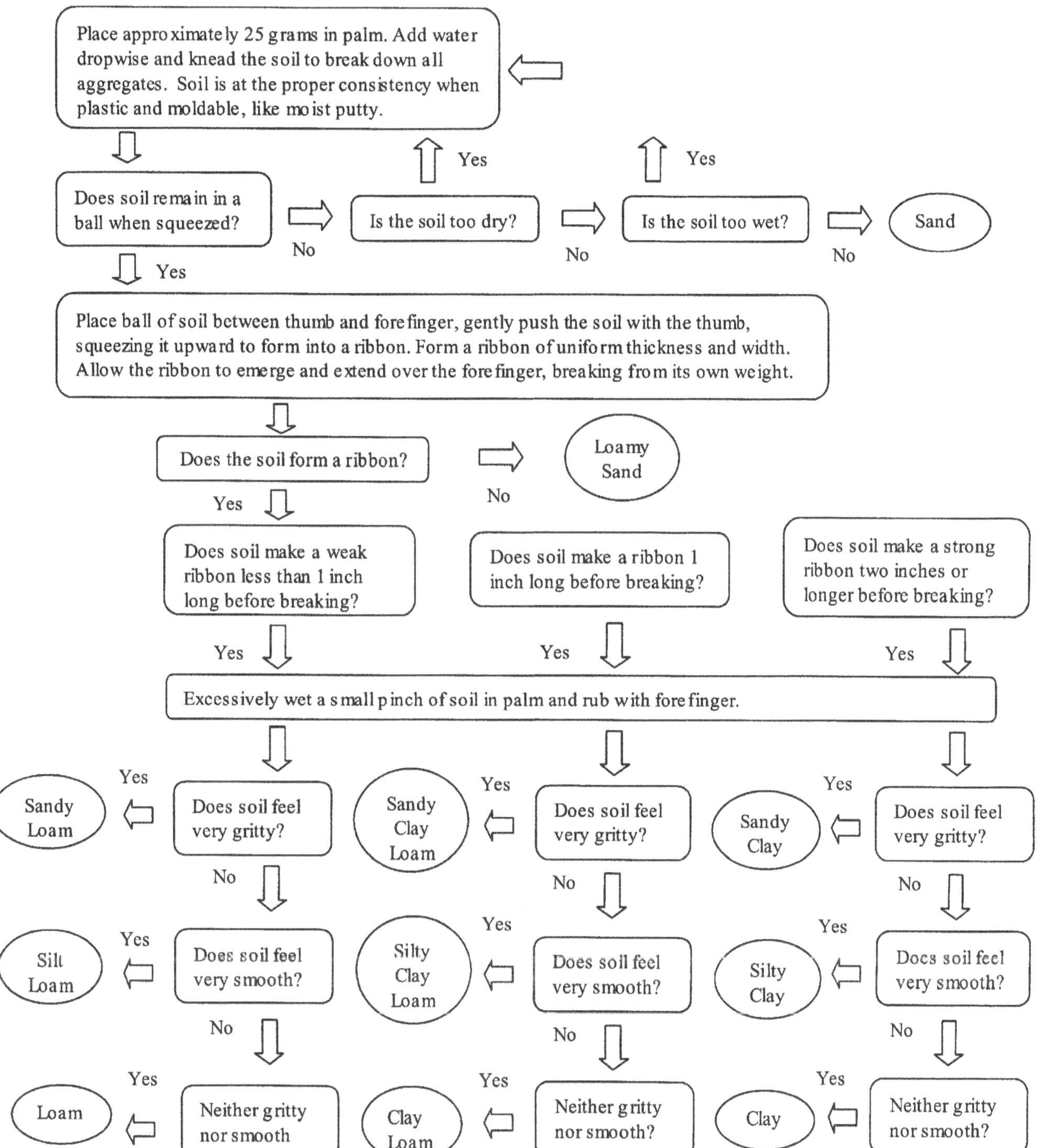

FIGURE 7.3 Flow chart for texture by feel method.

Soil drainage class can be estimated by integrating several **topographic** and **edaphic** factors. Soil scientists observe the site, its slope position and angle, the soil texture and depth, as well as any other features and place the site in a drainage category. The **parent material** can be also be observed if a soil pit is dug or if there are nearby road cuts or other areas to observe a cross-section of the soil. Geologic maps of the local area and county soil surveys can be used to help determine the parent material type. County soil surveys can provide a great deal of information on the character of the soil, as well as the use of the soil for various purposes such as what crops to grow, wildlife habitat values, wetland conditions, construction limitations and other considerations.

Soil temperature can be determined with a mercury bulb thermometer with the bulb covered with a preventive shield. Many thermocouple temperature devices are available as well. Soil temperature is an important environmental factor on its own but also as it interacts with soil moisture.

Soil pH is a relatively easily measured parameter that gives some general chemical characteristics of the soil that relate to nutrient availability, acidity, etc. It can be approximated by using easily obtained field soil pH kits or soil samples can be collected and taken to a laboratory where a pH meter can be used to obtain a more exact measure.

Soil moisture is an important soil feature in terms of defining the kinds and abundance of microorganisms, soil animals and plants. Elaborate methods exist to obtain exact measures of soil moisture at a given time or to monitor soil moisture over time. These methods are described in most soil science textbooks. For this study, we will be using an electronic soil moisture meter that uses probes that, when inserted into the soil, use conductance as a proxy for soil moisture at the time of measurement.

REFERENCES

Brady, N. C. 1974. *The Nature and Properties of Soils,* 8th Edition. MacMillan Publ. Co., Inc., New York, NY. 639 p.

Cox, G. W. 2002. *Laboratory Manual of General Ecology,* 8th Edition. McGraw-Hill, New York, NY. 320 p.

Harpstead, N. I., Sauerank, T. J. and W. F. Bennett. 1997. *Soil Science Simplified,* 3rd Edition. Iowa St. Univ. Press. Ames. 220 p.

Jenny, H. 1980. *The Soil Resource: Origin and Behavior.* Springer-Verlag, New York. 377 p. Milford, M. H. 1970. *Introduction to Soils and Soil Science-Laboratory Exercises.* Kendall Hunt Publishing Company, Dubuque, IA. 111 p.

Soil Survey Staff. 1951. *Soil Survey Manual.* U.S. Dept. Agr. Hdbk. No. 18. U.S. Gov't. Printing Office, Washington, D.C.

Soil Survey Staff. 1975. *Soil Taxonomy: A Basic System of Soil Classification for Making and Interpreting Soil Surveys.* U.S. Dept. Agr. Hdbk. No. 436. U.S. Govt. Printing Office, Washington, D.C. 754 p.

Assignment – Soil Characterization and Draft of Introduction

This assignment will be completed in three parts:

1. Following your TA's instruction, collect 3 soil samples of adequate size from each site (upland savanna, upland forest, and bottomland forest). Take each sample through the soil texture by feel procedure (Fig. 7.3) and record your conclusions. This will ultimately be included in the results section of your ecological report.

2. Write a draft your introduction. This should be typed, 12-pt font, 1-inch margins, and 1-2 pages long single-spaced. The introduction is where the rationale behind the study is first introduced (i.e. why are you doing this study?). Here you should give detailed background information regarding the park itself (e.g. historic uses, precipitation patterns, current use) and of the two vegetation types in which you will be sampling (defining characteristics of an upland savanna and bottomland forest).

3. Complete the questions below using information given to you by your TA during the trip, information given in the lab manual, and any helpful material you can find in the peer-reviewed literature.

   a. In which ecoregion is Lick Creek Park located? Of which major river is Lick Creek a tributary?

   b. What is the scientific name for Post Oak?

   c. What is the name (common or scientific) of the plant most often responsible for local hayfever problems?

   d. What is one of the most common grasses in the Blackland prairies?

   e. What was the past history of human use of the park?

f. What plant dominates the understory of the Post Oak Woodlands?

g. How can you tell a Blackjack Oak from a Post Oak?

h. What endangered species was Lick Creek Park set aside to protect?

i. How can Water Oak be distinguished from Post Oak or Blackjack Oak?

j. What is an epiphyte? Give an example that might be found in Lick Creek Park.

k. What are lichens? Give an example that might be found in Lick Creek Park.

l. What are the oxbow "lakes" ?

m. What is the dominant herbaceous species in the floodplain sedge meadows?

n. Why might it be easy to confuse a Water Oak with a Willow Oak?

o. How does the water level and flow of Lick Creek change seasonally?

p. How can Poison Ivy be distinguished from Virginia Creeper?

7.3 LCP Vegetative Sampling

The following design should be followed for both the savanna upland and bottomland forest sites.

Work in three groups: each group conducts sampling at four points along a given straight line from a designated starting point along a primary transect. Record all data using the Fulcrum app on the provided iPad.

1. Randomly generate one number depending on group (1st number). This number will give you your starting point along the primary transect.
   - Group 1: between 0 and 33
   - Group 2: between 34 and 67
   - Group 3: between 67 and 100
2. Locate your starting point (1st number) on the main transect.
3. Flip a coin. This will tell you if you are going left or right from your point
   - Heads: Left
   - Tails: Right
4. From your starting point, create a secondary transect perpendicular to the main transect of 60 meters in length. This transect will be your sampling transect.
5. Randomly generate 4 more numbers between 0 and 50. These numbers will correspond to your group's sampling points.

At each of the four points along your sampling transect:

Sample the herbaceous layer with the quadrat method: locate the 0.125 m^2 quadrat frame 3 meters to left (with your back to the primary transect) of the sampling point perpendicular to the line, record total canopy cover (all herbaceous vegetation) and then record species group(s) presence/absence in quadrat.

Sample the shrub/sapling layer with the line intercept method: lay a 10 meter tape starting 2 m to the right of the sampling point, again perpendicular to the line. Record the lengths of intercepts of woody species (less than 1.5m tall).

Sample the canopy layer with the point-center-quarter method: divide the area surrounding the point into 4 quadrants, using the transect to delineate the left from right halves and then a perpendicular line to delineate the front from back halves. Locate the nearest tree (dbh >= 10cm, live) in each quadrant, record the distance between the point and the tree, and the species group and DBH of that tree.

Based on the data submitted through Fulcrum, your TA will provide a spreadsheet with aggregate class results. You are responsible for presenting these (aggregate) results in a readable format in the results section of the ecological report.

Assignment – Draft of Methods; Revision of Introduction

This assignment will be completed in two parts:

1. Write the methods section of the report: describe the sampling procedures and methods used to sample the vegetative community. Avoid using lists and, instead, explain in detail each step performed in the vegetative portion of this study.

2. Based on the feedback from your TA, revise and re-write the introduction section of your report.

7.4. LCP Animal Community Sampling

The use of infrared-triggered activity monitors in ecological studies has recently increased. Commercial and non-commercial trail monitors have allowed biologists to learn more about animal behavior/biology and in facilitating population estimates. For active trail monitors, events are recorded when an object breaks the electrical pulse between the receiver and transmitter. Camera traps, or game cameras, are triggered by any movement (usually a threshold can be customized) within the field of view. Prior to this lab, your instructors placed 6 cameras (3 upland savanna, 3 bottomland) at LCP. Your lab assignment is to review photos taken from each of these cameras and identify species found in each of these sites.

Conduct the following:

> Each group will take one camera. Each person will view frames within 60 time series. Identify and number and type of species from each camera. Using the Fulcrum app on the iPads fill in all of the info for each data series (save after each time series). Do not forget to sync when finished.
>
> Compare results across groups/sites and use the aggregate class averages (provided by your TA) in your ecological report.

7.5. Completion of the Ecological Report

Complete your ecological report:

Revise the introduction and methods sections; respond to any feedback from your TA.

Write the results and discussion sections and submit a complete report

Results and Discussion section: Compare the upland savanna and bottomland forest communities based on your results. How do physical environment, plant community, and animal community differ among sites? State each important result and give an ecological explanation.

Save your report in a Microsoft Word file and submit via the course eCampus page.

Ecological Concepts and Modeling

Purpose

This final chapter will build on previously defined ecological concepts and discuss more broad ecological theory and application. The theory of islands biogeography and its implications for restoration are explored and biological modeling is introduced as a tool for progress and understanding. Populations of a species can be distributed in several ways across the landscape, including those that arranged into isolated groups and those that are linked through movements between groups. For multiple isolated populations, each population experiences a susceptibility to extinction without the possibility of natural recolonization. From a practical standpoint, we need to know if the separate populations are all likely to be impacted by the same environmental events, especially those that could cause a catastrophic decline in numbers such as a hurricane, fire, or disease. If the fates of extinction of isolated populations are independent, then the probability of simultaneous extinction of all populations would be much less than that of any single population. In contrast, if the fates of the populations are completely correlated, the probability of total extinction would be the same as for one population. Thus, from a restoration perspective, understanding the distribution of populations of a species in space, the internal dynamics of each population, and if movement between populations is occurring, are all core aspects of designing a plan for conserving a species.

8.1. Biogeography

Biogeography is a broad term and refers to the study of the geographic location of a species. This not only includes presence or occurrence, but also the movement (immigration and emigration) and colonization of species across spatial extents. **Island Biogeography**, then, is the study of establishing the mechanisms responsible for colonization, immigration, and emigration in an island system. In this context, an "island" is not just an area of land surrounded by water, but also refers to any environment with surroundings unsuitable for occupancy or use (e.g. patches of forest separated by roads or other development, lakes, mountain tops). This theory, coined by Robert MacArthur and E.O. Wilson in 1967, helps us to predict the number of species that would exist on a newly created island and explains how the relationship between the distance and area regulates **immigration** (the appearance of a new species) and **extinction** (the disappearance of a species).

The initial concept of a **metapopulation** structure has been revised and expanded as researchers have conducted field investigations of animal movements and genetics of **subpopulations**. The key contribution of the metapopulation concept to applied wildlife population biology is a better understanding of the importance of animal movements—dispersal—in management and conservation. Studies of dispersal have shown that it is not only the area that animals live in for extended periods of time (i.e., a biological season such as winter or summer) that are critical for survival, but also the areas that animals must travel through to reach areas of seasonal use that must be identified, studied, and allowed for to ensure survival. Discontinuities in environmental conditions occur as ecotones or relatively sharp breaks in environmental conditions, or as ecoclines or broader gradations in conditions over areas of greater geographic extent. Discontinuities can occur naturally, as with changes in soil type or edges of water bodies, or anthropogenically, as with agricultural lands or roads. The cause of the discontinuity is less important to wildlife occurrence and health than is the nature and extent of the discontinuity.

The **equilibrium diversity value** of the island is the species richness at which immigration balances extinction (Fig. 8.1). The equilibrium value depends on the

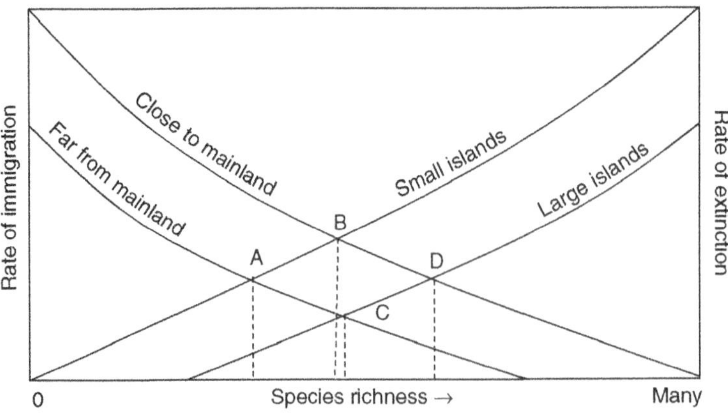

FIGURE 8.1 Graphical representation of island biogeography theory; Conceptual equilibrium diversity values for far, small islands (A), close, small islands (B), far, large islands (C), and close, large islands (D).

area of the island and the distance of that island to the mainland or source. In general, the **target effect** predicts that larger islands will have higher immigration rates (they are larger targets). Immigration rate is also predicted to be higher at closer islands than on islands further from the mainland or source. Additionally, the **rescue effect** predicts lower rates of extinction on closer islands due to the increased immigration and recolonization.

8.2. Restoration

The theory of island biogeography shapes many modern ecological concepts. An increasingly common term, **restoration** is, by definition, the act of restoring (bringing back into existence or use). The field of restoration encompasses two general schools of thought: **ecological restoration** is the application of restoration techniques or methods and **restoration ecology** is the science or theory of restoration. Successful restoration is guided by a sound understanding of the target species or system as well as the desired conditions to which the system is to be restored. From this, clear objectives for the restoration effort or project are outlined and techniques should be established.

Restoration Techniques

Restoration projects can be approached from multiple perspectives and employ a vast diversity of techniques. Introduction involves the **introduction** of species into an area in which they have not previously inhabited. **Reintroduction**, then, is the introduction of species into an area in which that have previously or historically inhabited. **Habitat fragmentation** is another term being used with increasing frequency in the ecological literature. It typically refers to the break-up, or addition of barriers to an area inhabited by a species (e.g. roads splitting patches of forest, urban development in a valley isolating two mountain tops). Fragmentation can be devastating especially for species particularly sensitive to human presence, only exist across a small/local spatial extent (**endemic**), or those that require a minimum area for breeding territory.

Fragmentation refers to the degree of **heterogeneity** of habitats across a landscape, with a focus on the isolation and size of resource patches available. Note that fragmentation is necessarily a species-specific condition. Various kinds or degrees of species-specific habitat heterogeneity can be described. In the extreme case, resource or vegetation patches can be isolated into islands surrounded by vastly different and, for specific species, unsuitable conditions. Even though a substantial decline in overall abundance might not be evident, partial isolation can effect population viability by incrementally lowering the numbers of animals per unit area of unsuitable and suitable environments in a landscape, lowering dispersal, and lowering the effective size of the breeding populations. Population viability is usually enhanced when individuals are able to move to between populations as within a metapopulation structure. At some spatial scale, all locations are heterogeneous (or patchy). What may appear to be homogeneous to a large mammal may be quite heterogeneous at the spatial scale of the amphibian. The connectivity among patches will often determine if a species can survive in a specific location and ultimately within a larger region. The term connectivity refers to the extent to which a species or population can move among landscape elements in a mosaic of habitat. This definition necessitates linkages among species at appropriate spatial and temporal scales. Corridors are one means of achieving connectivity. Until recently most species lived in well-connected landscapes. However, human-induced impacts can disrupt the natural connectivity. A narrow path in the woods might prevent an amphibian from crossing, a condition that would have little impact on the movements of a larger and more mobile animal. Thus, the task of the restorationist is complicated by the species or group specific abilities of animals to move across potential barriers. A restoration plan should include a categorization of species by the potential impacts that various paths, roads, structures, changes in vegetation structure, and other features of the plan might have on species of interest. As such, many conservation biologists have been advocating the retention or development of corridors that help to link landscape patches. The intuitive appeal of corridors resulted in the widespread recommendation for their use in conservation planning and landscape design. But care must be taken to ensure that the species of interest actually use these corridors.

In this way, we can think of this approach to restoration as an application of island biogeography theory where the environmental patches represent islands.

8.3. Ecological Modeling

Changes in population size over time, that results from interactions with the physical environment and with other biological populations, is referred to as population dynamics. We can repeatedly sample a population and its biophysical environment over time to evaluate the dynamics of the population and the interactions. Alternatively, we can also use models to study population dynamics. A model is a simplified representation of a real system. Modeling, the development, and application of models, can improve understanding of ecological systems, predict behavior of ecological system, and test ecological hypothesis. Once developed and tested, models can be used to conduct "what if" experiments, especially experiments that are infeasible or impractical to conduct physically (e.g., global warming, environmental impact assessment), and optimize environmental decision-making. Modeling can also help identify knowledge gaps and formulate testable hypothesis.

The simplest population growth model is the **exponential growth model**. When resources (food, cover, reproduction sites, etc.) are not limiting, population growth is determined by simply the present population size (# of individuals) and the growth rate:

$$Nt + 1 = Nt + r\,Nt$$

Where Nt and Nt + 1 are the population size at time t and t + 1, respectively; and r is the (intrinsic) growth rate, population growth per time unit per reproductive individual, determined by birth rate minus death rate.

Resource (food, cover, reproductive sites, etc.) will, however, become limited if population continues to grow. A given environment may support only a certain number of species. The population size that an area has the resources to support is called the carrying capacity (K), which was discussed in a previous chapter. The exponential growth model can be modified to form a **logistic growth model** based on the idea of carrying capacity:

$$Nt + 1 - Nt = Nt\,r\,(1 - Nt/K)$$

When the population size is very small (Nt close to 0), population will grow almost exponentially; with increasing Nt (or $1 - Nt/K < 1$), population growth will decrease, and eventually stop growth when $Nt = K$ (or $1 - Nt/K = 0$).

Interactions with populations of other species can also affect the dynamics of a population. Predator-prey interaction is a particularly interesting interspecific (between species) interaction that has cyclic dynamics produced by a feedback mechanism. An increase in the predator population reduces the prey population, the reduced prey population (food) leads to a decline in the predator population, the prey population recovers with reduced predation, which provides the resource and allows the predator population to increase.

REFERENCES

Gotelli, N. J. 2001. *A Primer of Ecology,* 3rd Edition. 265 p.

Grant, W. E., Pedersen, E. K. and S. L. Marín. 1997. *Ecology and Natural Resource Management: Systems Analysis and Simulation.* Wiley, New York. 373 pp.

Holt, R., Lawton J., Polis G. and N. Martinez. 1999. Trophic rank and the species-area relationship. *Ecology*. 80:1495–1504

Morrison, M. L. 2009. *Restoring Wildlife: Ecological Concepts and Practical Applications*. Island Press.

Assignment: Patchy Prairies Simulation

This lab exercise will be conducted using the SimBio software available through SimuText.

You will first use the information provided to you in the registration email to create a student account and then will have access to the simulation models and associated assignment questions.

Lab Objectives

The purpose of this lab exercise is to become familiar with basic biological modeling as well as the restoration concepts introduced earlier in the chapter. The software will guide you through a series of scenarios in which you will use your knowledge of island biogeography theory, modeling, and ecological restoration to balance the needs of an endangered butterfly and development.

Following the exercise, please answer the associated questions provided by SimBio and turn into your TA.

Appendix

Conversion Table

Length

inches (in.) × 2.54 = centimeters (cm)
feet (ft) × 0.3048 = meters (m)
yards (yd) × 0.9144 = meters (m)
miles (mi) × 1.6094 = kilometers (km)
statute miles × 0.8684 = Nautical miles

centimeters (cm) × 0.3937 = inches (in.)
meters (m) × 3.2808 = feet (ft)
meters (m) × 1.0936 = yards (yd)
kilometers (km) × 0.6214 = miles (mi)
nautical miles × 1.15 = statute miles (mi)

statute miles (mi) × 80 = chains
statute miles (mi) × 320 = rods
statute miles (mi) × 5280 = feet (ft)
rods × 16.5 = feet (ft)

Area

square inches (in.^{2}) × 6.45 = sq centimeters (cm^2)
square feet (ft^2) × 0.0929 = sq meters (m^2)
square yards (yd^2) × 0.8361 = sq meters (m^2)
square miles (mi^2) × 2.5900 = sq kilometers (km^2)
acres (a) × 0.4047 = hectares (ha)

square centimeters (cm^2) × 0.155 = sq inches ($in.^2$)
square meters (m^2) × 10.7639 = sq feet (ft^2)
square meters (m^2) × 1.1960 = sq yards (yd^2)
square kilometers (km^2) × 0.3831 = sq miles (mi^2)
hectares (ha) × 2.4710 = acres (a)

acres (a) × 43,560 = sq feet (ft^2)
hectares (ha) × 10,000 = sq meters (m^2)

Volume

cubic inches ($in.^3$) × 16.39 = cubic cm (cm^3)
cubic feet (ft^2) × 0.028 = cubic meters (m^3)
cubic yards (yd^2) × 0.765 = cubic meters (m^3)
cubic miles (mi^2) × 4.17 = cubic kilometers (km^3)

cubic cm (cm^3) × 0.06 = cubic inches ($in.^3$)
cubic meters (m^3) × 35.30 = cubic feet (ft^2)
cubic meters (m^3) × 1.3079 = cubic yards (yd^2)
cubic kilometers (km^3) × 0.24 = cubic miles (mi^2)

Liquid Volume

ounces (oz) × 30 = milliliters (mL)
pints (pt) × 0.47 = liters (L)
quarts (qt) × 0.95 = liters (L)
gallons (gal) × 3.8 = liters (L)

milliliters (mL) × 0.034 = ounces (oz)
liters (L) × 2.1 = pints (pt)
liters (L) × 1.06 = quarts (qt)
liters (L) × 0.26 = gallons (gal)

Mass and Weight

ounces (oz) × 28.3495 = grams (g)
pounds (lb) × 0.4536 = kilograms (kg)
short tons (tn) × 0.91 = metric tons (t)

grams (g) × 0.03527 = ounces (oz)
kilograms (kg) × 2.2046 = pounds (lb)
metric tons (t) × 1.10 = short tons (tn)

Density

pounds/ft^3 (lb/ft^3) × 16.0185 = kilograms/m^3 (kg/m^3)
pounds/gallon (lb/gal) × 119.826 = kilograms/m^3 (kg/$m^3$3)

kilograms/m^3 (kg/m^3) × 0.06243 = pounds/ft^3 (lb/ft^3)
kilograms/m^3 (kg/m^3) × 0.008345 = pounds/gal (lb/gal)
kilograms/m^3 (kg/m^3) × 1 = grams/liter (g/L)

Velocity

miles/hour (mph) × 0.448 = meters/second (mps)
miles/hour (mph) × 1.6094 = kilometers/hour (kmph)
miles/hour (mph) × 0.8684 = knots (kn) (nautical mph)

meters/second (mps) × 2.24 = miles/hour (mph)
kilometers/hour (kmph) × 0.62 = miles/hour (mph)
knots (kn) (nautical mph) × 1.15 = miles/hour (mph)

Temperature

Fahrenheit (°F) –32 × 5/9 = Celsius (°C)
Celsius (°C) × 9/5 + 32 = Fahrenheit (°F)

Energy

joule × 0.2391 = calorie (cal)
joule × 1 = watt-second (W-sec)
calorie (cal) × 4.1819 = joule
cal/cm^2/min × 697.8 = watt/m^2

Illumination

foot-candles ≡ 1 lumens/sq feet
lux ≡ 1 lumens/sq meter
foot-candles × 10.764 = lux
lux × 0.0929 = foot-candle